Sonic Intimacy

THE STUDY OF SOUND

Editor: Michael Bull

Each book in The Study of Sound *offers a concise look at a single concept within the field of sound studies. With an emphasis on the interdisciplinary nature of the topics at hand, the series explores a range of core issues, debates and objects within sound studies from a variety of perspectives and within a multitude of contexts.*

Editorial Board:

Carolyn Birdsall, Assistant Professor of Television and Cross-Media Culture, University of Amsterdam, The Netherlands

Martin Daughtry, Assistant Professor of Music, Arts and Humanities, NYU, USA

Michael Heller, Associate Professor, Department of Music, University of Pittsburgh, USA

Brian Kane, Associate Professor, Department of Music, Yale University, USA

Marie Thompson, Lecturer, School of Film and Media, University of Lincoln, UK

James Mansell, Assistant Professor of Cultural Studies, Department of Culture, Film and Media, University of Nottingham, UK

Published Titles:
The Sound of Nonsense by Richard Elliott
Humming by Suk-Jun Kim
Lipsynching by Merrie Snell
Sonic Fiction by Holger Schulze
Sirens by Michael Bull
Sonic Intimacy by Malcolm James

Forthcoming Titles:
Wild Sound by Michael Pigott

Sonic Intimacy

Reggae Sound Systems, Jungle Pirate Radio and Grime YouTube Music Videos

Malcolm James

BLOOMSBURY ACADEMIC
NEW YORK • LONDON • OXFORD • NEW DELHI • SYDNEY

BLOOMSBURY ACADEMIC
Bloomsbury Publishing Inc
1385 Broadway, New York, NY 10018, USA
50 Bedford Square, London, WC1B 3DP, UK

BLOOMSBURY, BLOOMSBURY ACADEMIC and the Diana logo
are trademarks of Bloomsbury Publishing Plc

First published in the United States of America 2021

Copyright © Malcolm James, 2021

For legal purposes the Acknowledgements on p. viii constitute
an extension of this copyright page.

Cover design and cover image by Liron Gilenberg | www.ironicitalics.com

All rights reserved. No part of this publication may be reproduced or transmitted in any form or by any means, electronic or mechanical, including photocopying, recording, or any information storage or retrieval system, without prior permission in writing from the publishers.

Bloomsbury Publishing Inc does not have any control over, or responsibility for, any third-party websites referred to or in this book. All internet addresses given in this book were correct at the time of going to press. The author and publisher regret any inconvenience caused if addresses have changed or sites have ceased to exist, but can accept no responsibility for any such changes.

Library of Congress Cataloging-in-Publication Data
Names: James, Malcolm (T. Malcolm) author.
Title: Sonic intimacy : reggae sound systems, jungle pirate radio and grime Youtube music videos / Malcolm James.
Description: New York : Bloomsbury Academic, 2020. | Series: The study of sound | Includes bibliographical references and index. | Summary: "An exploration of sonic intimacy, how it has transformed in relation to music technologies, and why this matters"– Provided by publisher.
Identifiers: LCCN 2020015537 | ISBN 9781501320712 (hardback) | ISBN 9781501320729 (paperback) | ISBN 9781501320736 (pdf) | ISBN 9781501320743 (ebook)
Subjects: LCSH: Reggae music–Social aspects–Great Britain. | Jungle (Music)–Social aspects–Great Britain. | Grime (Music)–Social aspects–Great Britain. | Pirate radio broadcasting–Social aspects–Great Britain. | YouTube (Electronic resource)–Social aspects–Great Britain. | Intimacy (Psychology)
Classification: LCC ML3917.G7 J36 2020 | DDC 306.4/84240941–dc23
LC record available at https://lccn.loc.gov/2020015537

ISBN: HB: 978-1-5013-2071-2
PB: 978-1-5013-2072-9
ePDF: 978-1-5013-2073-6
eBook: 978-1-5013-2074-3

Series: The Study of Sound

Typeset by Integra Software Services Pvt. Ltd.

To find out more about our authors and books visit www.bloomsbury.com
and sign up for our newsletters.

For the ravers.

CONTENTS

Acknowledgements viii

1 Introduction 1
2 The reggae sound system and vibe 27
3 Jungle pirate radio and hype 55
4 Grime and YouTube music videos 81
5 Conclusion: From left critique to alternative cultural politics 105

References 120
Index 136

ACKNOWLEDGEMENTS

This book has been made possible by the generous support of some important and much-loved people. I would like firstly to thank Michael Bull for suggesting that I write for this series, and for continually reminding me (and everyone) of the joys of intellectual pursuit. I would like to thank Sivamohan Valluvan for his warm guidance, ever-prescient commentary on, and close reading of, a number of themes and ideas in the book – many of which were seeded in our conversations. Deepest thanks to Helen Kim for helping me improve the Introduction; and to Helen, Antonia Dawes and Nabila Munawar for comradeship (scholarly and otherwise), and for helping me articulate my ideas on diaspora and intimacy through our performative work on the theme. It was a pleasure to discuss ideas for Chapter 2 with Martin Evans, and to find there a wonderful colleague. Johan Koelb generously introduced me to the PALM archive and the Duke Vin interviews held there, and my discussion of reggae benefits immeasurably from that. Caspar Melville and I shared a number of determining conversations on jungle and associated genres. Chapter 3 is indebted to his capacious knowledge of music. In Tom Cordell I have found a kindred spirit and expansive autodidact. Our long debates on jungle, pirate radio, capitalism and other subjects unavoidably shaped this book. The same is true of Alberto Duman who has over the last years shown me the pleasure of unstinting, genuinely interdisciplinary and bottom-up collaboration. Original junglist Christine Clark provided brilliant insight into the British–Dutch jungle nexus, and original acid technician Pete Watkins shared his vast knowledge of dance music technologies. Without either, my understanding of jungle would have been all the poorer. Malacki Doyley-Williams and Jordan King shared their profound thoughts on grime, music, society and YouTube. Chapter 4 owes much to their sharp analysis and observation. Joy White provided essential feedback on the same chapter and, with Dan Hancox, has been a source of inspiration on the subject matter of grime. I would like to thank Matthew Hayes for his companionship and for so kindly hosting me at St Thomas, Canada, to talk about this subject and to Katie Thorsteinson who provided such an insightful comment after that talk that I partially re-wrote the introduction of the book. Baljit Bains continually challenges me to think better, and in the case of this book encouraged a much welcome reappraisal of Ralph Ellison's 'Living with Music'. Naaz Rashid, who is so important, not just to my academic endeavours, kindly helped me rework the cover copy such that it made sense to someone other than me. To my comrades in life and at my branch of the University and College Union (UCU) my deepest gratitude and solidarity. Amidst the fuckery, your steadfast soulfulness and socialism makes all the difference. And finally, to Nicole and Sonny, my parents, sister and family for all the love. None of this is possible without you!

1

Introduction

In April 2017 I went to see Nadia Rose play at Village Underground in Shoreditch, London. I'd previously seen her videos on YouTube and was impressed by her affinity with the camera. I tried to buy a physical copy of her EP but it was only available on YouTube. It was later released on MP3. Wanting to support the artist, I paid to download the tracks. When I did I was then struck by what they lacked. I had come to know Nadia Rose and her music almost exclusively through YouTube. I was expecting the audio to reproduce its effects. It didn't.

In a third-year undergraduate class on post-colonial approaches to the city, I encouraged students to think about what these shifts might mean for music, and indeed for the alternative cultural and political registers of black diasporic sound culture. We went back through Debord, Benjamin, Adorno and Wynter, thinking about the relationship between visual culture, capitalism and race, and then engaged with Gilroy, Du Bois and Weheliye to explore the alternative relation of sound to modernity.

Back at Village Underground, Nadia Rose's Croydon crew had turned up, but the rest of the audience was only there on the back of the Skwod video. They didn't know the other tracks. Aloof and half-interested, as London crowds can be, they were draining, and Rose, who is so confident on screen, has such proficiency in the medium of her generation, was perceptibly nervous. Granted, this was a big gig, but she also seemed unfamiliar with these live arts. A graduate of Brit School, clearly trained to a high level in video performance, her YouTube persona was not easily communicated to the crowd. The little gestures, flicks, facial movements that are caught on camera to such great effect seemed lost in the physical distance between her body and the dead weight of the mêlée. Her greatest moments of comfort appeared to come when her 'skwod' gathered on stage, evidently friends, and she could feel the intimate embrace of the cypher. Her other moment of comfort came as she physically abandoned those assembled in front of her to live Snapchat from stage to a screen following that she said gives her love.

This book's enquiry on the transformation of sonic intimacy and alternative cultural politics lies between this moment (the interface of black diasporic sound culture with social media) and two others: the reggae and dub sound systems of the 1970s and 1980s and jungle pirate radio of the 1990s. Half a century prior to Nadia Rose's appearance at Village Underground, hundreds of reggae and dub sound systems were producing their bass-mediated demands in excess of the racist state of the moment (Gilroy 1987). As captured in Franco Rosso's film *Babylon*, their low-end properties collected the dance floor and a largely black and working-class contingent, while providing the raw material to project an alternative imagination of Britain (Rosso 1981). A decade or so later, hundreds of jungle pirate radio stations blasted rolling bass lines across British cities moving under the pressure of two decades of Tory rule. Between bedrooms and cars, working-class and multi-ethnic dialogues were sustained.

Between these three moments (reggae sound systems, jungle pirate radio and grime YouTube music videos) lies an important question about the transformation of alternative black diasporic sound culture in Britain. How did the bass-mediated demands of the reggae and dub sound system bleed into the fractured fervour of pirate radio in the 1990s and then into the hyperlinked intensities and immediacies of YouTube music videos from 2008 onwards? To what extent was the tactility and proximity of the sound system necessary for the kinds of alternative politics it generated against the racist state? How can these politics be evaluated in relation to an illegal radio infrastructure, dispersed, extensive and fully charged with the agonism of John Major's Britain? And how should both be considered in relation to YouTube music videos' corporate ownership, atomization and digital screen intensities?

The book addresses these questions through lending an empathetic ear to modernity's relational flows of alternative and dominant formations. That is also to say that it does not answer them through an argument of linear decline – the sound system losing out progressively to jungle pirate radio and grime YouTube music videos. Attractive as the 'capitalism as progress, capitalism as ever and all consuming' thesis is to certain branches of Marxist media scholarship, what cultural theory doesn't need is another negative restatement of myth-mantra of dominant capitalism as modernity.

That such questions are concerned with *sonic intimacy* might not be immediately apparent. After all, the notion of *sonic intimacy*, while habitually known, is also somewhat illusory. Scholarship on the sonic has opened up regimes of knowledge not fully captured by the visual, and literature on intimacy has enabled an exploration of relationships, inter-subjectivity, feeling, forces and vibes not present in textual approaches to social and cultural life. But the relation of the sonic and intimate to alternative cultural politics has infrequently been considered. Rather, with the exception of some select works addressed below, the vibe of a dancehall or the hype of pirate radio has tended to provide the atmosphere for scholarly discussion, rather than be the focus. This is not inconsequential for our understanding of alternative cultural politics in late modern society, especially if we consider that black diasporic sounds and their intimacies have had a historically close relationship to the alternative forms of being and knowledge practised, produced and projected by peoples marginalized by the textual, visual and racial regimes of Eurocentric capitalist modernity (Hall 1997; Wynter and McKittrick 2015).

The remainder of this chapter clarifies this assertion, opening up the dynamics of sonic intimacy and locating them in the conjoint fields of race, post-colonial, sound and cultural studies. The remainder of the book is divided into three chapters, each focusing on a different sound culture and conjuncture. Chapter 2 explores the 1970s and 1980s reggae and dub sound system, Chapter 3 addresses 1990s jungle pirate radio, and Chapter 4 post-2008 grime YouTube music videos. This conjunctural approach does not imply that these moments are discrete. Indeed, the book is organized so as to draw through their continuities and discontinuities. But it does also imply that reggae sound systems, jungle pirate radio and grime YouTube music videos can be understood as particular clusters of sonic, social and technological materials and that these clusters are archetypical of their moments. Understanding them as such then helps to identify what is at stake in sonic intimacy's modern remaking.[1]

The sonic

'Bass history is a moving/is a hurting black story' intones dub poet Linton Kwesi Johnson on the 1980 album *Bass Culture* (Johnson 1980). In that landmark collaboration with Dennis Bovell, Johnson speaks poetic verse through dub's deconstructive rhythmatics, drawing attention to the work of sound, and bass, in conveying the history of black Atlantic people in Britain. Eight decades earlier, US sociologist W. E. B. Du Bois made similar comment on black US culture. In his seminal work *The Souls of Black Folk*, slave spirituals function as central allegories for an exploration of racialized being and knowledge. Those haunting melodies welled from the black souls of America's dark past. Scored in minor key, they 'articulate[d] the message of the slave to the world' (Du Bois 2007: 3–4, 169) – a message not intelligible to slave-owning households. For Plantation Lady Mary Boykin Chesunt 'the words have no meaning at all'. For her, the handclaps, 'shrill shrieks', a 'minor key' 'was all sound' (cited in Shaw 2013: 138). Indeed, like the reggae bass line, many of those songs did not have literal meaning, but they carried with them knowledge both terrible and joyous. They communicated a message to the world, 'a hope – a faith in the ultimate justice of things' (Du Bois 2007: 169, 175).[2]

---

[1]The three sound cultures selected for this book are not chosen to be representative of British diasporic sound and music cultures in the post-war period, or of all associated sonic intimacies. Rather, they are selected because of their prominent place in popular culture and because of their particular affinities with their moments. These affinities, then, provide the basis for making a theoretical argument about the transformation of sonic intimacy in late modernity. Readers interested in the interface with the black diasporic and south Asian sound culture are advised to engage additionally with some of the excellent work in the field (Sharma et al. 1996; Kim 2012; 2015).

[2]'The child sang it to his children and they to their children's children, and so two hundred years it has travelled down to us and we sing it to our children, knowing as little as our fathers what its words may mean, but knowing well the meaning of its music' (Du Bois 2007: 170).

The sound cultures discussed in this book contain reverberations of those moments, but reggae sound systems, jungle pirate radio and grime YouTube music videos are not slave spirituals just as sure as a reggae bass line is not a jungle one. Of different times, produced and productive in different moments of capitalism, racism and colonialism, they are part of each other's stories as they are resolutely of their own. And still, to differing degrees, they convey through sound alternative stories of modernity, and to that end Linton Kwesi Johnson and Du Bois's remain prescient.

After all, the over-determination of being and knowledge by visual and textual means has remained a central plank of modern Western societies, as has the coterminous development of the visual and textual with racism and capitalism (Dyer 1997; Wynter and McKittrick 2015). In the margins of these dominant articulations of modernity, black diasporic sound cultures have continued to provide alternative playlists for the modern condition, just as they have offered registers through which demands beyond those positions have materialized. It is for this reason that the reggae sound system is written about as:

> ... the recorded documentation of an alterative living history of the black presence in Britain; a space in which black Britons could tell, intellectualise and become conscious of social justice, black rights, anti-racism and against police harassment with people who understood and had sympathy with it. (Henry 2006: 8)

The use of 'alternative' rather than 'oppositional' in the above quote is instructive. For Henry, black and working-class reggae sound systems are culturally alternative to the dominant white petit-bourgeois culture of the day. They are alternative, not neatly oppositional because they contain elements of dominant culture, not least pieces of consumer technology.[3]

For Henry, the UK reggae sound system documents an alternative living history to formal written culture and to the white bourgeois

---

[3]In *Marxism and Literature*, Williams discusses the relation of dominant culture to any conjuncture, but also how in that same conjuncture other forms of culture exist that are alternative and oppositional to those dominant forms (1977: 121–7).

literacy of the enlightenment, precisely because of its marginal relation to it. Its documentation is not primarily found in the textbook (although there is now a significant literature in this area) but in its accumulated traditions, expressions and politics, many of which are sonic.

This is not to make the case for the neat separation of sound and text, and by inference blackness and whiteness. Such binary conflations are in fact unhelpful in thinking through the transformation of sonic intimacy through modernity. The sonic and symbolic remain intertwined, both are part of the racial structuring of modernity, just as black diasporic sound cultures resist, but can also reproduce dominant social orders.

To draw out these textures and tensions, this book addresses sound not as a 'thing' or an 'effect', but as relation. For that reason sound is discussed as 'sonics'. The *sonic* is not then synonymous with 'sound'. Rather it is concerned with the ways in which sound is culturally formatted (Wicke 2016: 26). From its earliest uses in Latin as *sonos*, the sonic has included both the structured materiality of sound and the hearing subject, which becomes a feeling subject when vibration is taken into account. In this way, the sonic relates sound to the subject and by extension to their everyday hearing/feeling contexts and social organizations, that is to say, to the cultural. The sonic is culturized sound matter.

In modern scholarship the visual has tended to take primacy over the sonic – whether that be sound, vibration, music or noise. It has been the visual, not the sonic, that has dominated discussions of twentieth-century modern culture. From advertising hoardings and cinema projections to flat screens, modern society has been saturated by the visual. That, as documented in Guy Debord's *Society of the Spectacle*, has profoundly affected how we *know* and how we *are*, pushing sound (which is as ubiquitous as light) to more marginal sites of public knowledge (Debord 2002).

Understanding the sonic in modernity is, then, fraught; the quality of sound not easily grasped by scholarly disciplines born of modernity and predisposed to (and often preoccupied with) the visual. But to comment on a preoccupation with the visual is not only to note an analytic predisposition for culturally informed arrangements of light. In fact, preoccupation with the visual is more particular. Whether it be the advertising hoarding or indeed the page of a book, the preoccupation has been particularly with the textual – which is to say, it has been concerned with *logos* – signs, signification and the like. In cultural studies this influence is ubiquitous. From the semiotics of Barthes and the textuality of Butler, to Hall's work on representation and Dick Hebdige's statement that youth subcultures (punk in particular) were 'pregnant with significance' (Hebdige 1979), it has been a textual reading of the visual that has provided the codex to decrypt the inner meaning of the modern. Indeed, this approach to culture has informed accounts on sound and music in everyday life (DeNora 2000). (This predilection for the visual-textual has not been helped by the conjoint (patriarchal) assignation of the ear as a vulnerable organ; an organ that could be penetrated; that could not be closed like the eye; that could not be mastered to attend critically to the dangers of modern life (Boutin 2015: 13–14; McLuhan 1967: 32; Simmel and Levine 1971)).

This is not to say that the relation of sound to modernity has gone entirely unnoticed. Scholars most associated with textual readings of modern life have also explored modernity through sound. Benjamin and Lefebvre both developed personal and poetic reflections on the sounds of cities. Benjamin's writing on Freiburg, Marseille and Berlin shows how the city's echoes add depth and alternative temporalities to the urban spectacle. His essay on Naples explores the porous rhythms of architecture, society, religion, history, emotion, eroticism, touch, smell and sound. Sonics spill from his pages as they overflow from crowded dwellings and communal streets (Benjamin 1979). Lefebvre's work – the last text before he died in 1992 – attended to the rhythms of Mediterranean towns. Positing a flâneur figure (the 'rhythmanalyst'), Lefebvre revealed his sonic sensitivity, one attuned more 'to times than to spaces, to moods than to images, to the atmosphere than to particular events' (Lefebvre and Regulier 2004: 87–8). Lefebvre's 'rhythmanalyst' was alive to the diasporic formation of these towns (part of their maritime history), and to the ways in which each peculiar locale develops its own reverberation.

As Benjamin rejected the narratives of progress and containable clarity that dominate bourgeois textual accounts of the city, Lefebvre's rhythmanalyst deployed a relational and distinctly post-colonial ear to the same end. Benjamin's anti-enlightenment sensibilities and Lefebvre's diasporic empathy form, then, a productive interface with black diasporic and post-colonial scholarship, but this is not to say that they were engaged with questions of race or colonialism in particular.

A largely unremarked upon dialogue between Fredric Jameson's *Postmodernism* (Jameson 1991) and Paul Gilroy's *Black Atlantic* (Gilroy 1993) helps us understand what is at stake in this partial absence. In those seminal texts, Jameson and Gilroy address similar epistemic moments, both are persistent in their call to engage with the doubling of modernity and both further concern themselves with the shifts from soul to commodification. For Jameson the soul of the modern becomes fractured through late modernity. For Gilroy, counter-modernities occur alongside the uneven capture of black diasporic demands by Eurocentric capitalism. And it is through these similarities that key differences in the texts become pronounced. Jameson's analysis is principally visual and aesthetic and omits the relationship of late modernity to racism, slavery and capitalism. For Gilroy, racism, slavery and capitalism are central to any consideration of the modern, the visual and the coded, and this leads inexorably to an enquiry into the sonic as a central modality through which black people have communicated themselves to the world in the context of visual and racial over-determination. When race and modernity are centred in the analysis, the sonic then 'supplement[s] and partially displace[s] concerns with textually', while also showing black diasporic sound cultures to be thoroughly modern (Gilroy 1993: 36).

This point, on the relationship between sound and modernity, is further elaborated through attention to technology. The significance of this move is greater than simply acknowledging technology's place in social and cultural life. Rather, bringing technology into the discussion further locates black diasporic sound culture in the modern, addressing thereby the racial over-investments we make in black diasporic sound culture when technology is ignored. Figuring modern technology in black diasporic sonics moves us beyond the 'zero-sum game of authenticity versus commodification' (Weheliye 2005: 48). Addressing sound culture *and* technology contests sedimented and romanticized discussions of sound and music. When reggae, for example, is discussed only in terms of sound, it can be figured as a pure and authentic black expression outside modernity, and we find similar positions taken on the blues, jazz and hip hop. This has the same stamp as the more obviously problematic argument that black music is primitive, intuitive and unmodern (Weheliye 2005: 48). Thinking about black music in technological terms – its instruments, studios, recording processes, distribution etc. – means thinking about it as modern music.[4]

This relation between modernity, technology and black diasporic music is explored in the final scene of Ralph Ellison's *Invisible Man*. Here Ellison's 'poetics [are] situated at the interstices of music and technology, [bridging] the assumed divide between black cultural production and modern informational technologies, probing their textual and overdetermined interdependencies rather than their contrasting natures' (Weheliye 2005: 47). This scene is discussed at length by Weheliye, but is also elaborated here to clarify Ellison's contribution to the *sonic*, and to lay the ground for a later discussion of Ellison's less observed, and not unproblematic, analysis of *intimacy*.

In the prologue to Ellison's eponymously titled book, the Invisible Man hibernates in a dark basement-cum-hole under a 'whites only building'. Having lived through multiple racial institutions – the industrial, the activist socialist and the street, and having been thwarted in his search for full humanity at each turn, the Invisible Man's final sojourn is to the underground. The scene in the basement of the whites' only apartment building allegorically captures the enduring inescapabilty of racial structure, just as it signifies the way in which the Invisible Man comes to know life beyond racialized being. In the darkness of the basement transcendental possibilities arise because the Invisible Man can finally see the visual regimes of racism that have operated on him, and through which he has been living. This is stressed by the inclusion of sound in the scene, which both clarifies racism's relationship to the visual and offers an alternative register for life not captured by it. Louis Armstrong's rendition of Fats Waller's protest song *Black and Blue* provides that soundtrack. The source of the sound that fills the basement is a phonograph. Like the light bulbs that also populate the basement, the phonograph is a quintessentially modern technology (Ellison 2001).

In this way, Ellison comments on the relationship between sound, the visual, technology and the modern Black American experience. The phonograph highlights the relation of the protagonist to sound recording and reproduction technologies, just as it does to the

---

[4]Jazz is as modern and American as the 'the automobile and the airplane', says Duke Ellington (Ellington 1993: 253, see also 80, 151).

modern and racial regimes of seeing and being seen. But for Ellison the Invisible Man never becomes a body among others. While he is displaced through sound, moving in the breaks of Armstrong's music, moving beneath the vibrations of the record player, it is still human life that is at stake. The speaker's vibration of the airwaves and the Invisible Man's eardrums exceed signification as they make the ending of technology and the beginning of the human difficult to define,[5] but the human is nonetheless what is politically at stake.

Intimacy

Intimacy is less familiar to cultural studies than scholarship on sound and the sonic, although it is well developed in English literature, psychology and psychoanalysis, and to a lesser extent in sociology. Much of this, of course, deals with love, sexuality and romance. And while the dynamics of these relations are relevant to this book, the forms of intimacy discussed in the following pages have more in common with intimacy's other applications: an intimate room, an intimate sound, an intimate feeling, an intimate relation, intimate knowledge and, indeed, an intimate atmosphere. As with the Latin *intimus*, the superlative of intimate, intimacy in this book is concerned with inner*most* depth, presence, wholeness and privileged knowledge, always in relation (Sexton and Sexton 1982: 2).

Debates on intimacy in the West are often staged over a pre-romantic/romantic/modern timeline corresponding to the ways in which intimacy with God became an infatuation with the soul of another, and then a project of self-betterment (Giddens 1992: 2; Illouz 2012: 39; Zeldin 1994: 325). The subjects of these debates are routinely European, with the Romantic period being particularly prominent in literary analysis (Yousef 2013).

In these Eurocentric accounts, property, family and the individual possession of *man's* own person are developed as sites of intimacy and cornerstones of European and Western liberty (Lowe 2015: 28). The intimate rooms, objects and interior (personal) musings that redound in romantic literature are not casual decorations. They are rather the privileged sites to which the bourgeois

---

[5]See also for discussion Sedgwick and Frank 1995: 8, 10.

individual withdraws from an alienating world, and, at the same time, the sites of the reproduction of bourgeois power. Bourgeois intimacy concerns the reproduction of property at the same time as it is contingent with reproduction of the bourgeoisie itself. The sexual intimacy of the European bourgeoisie, Boym writes, took place surrounded by their 'innumerable curio cabinets and chests of drawers' (Boym 2001: 255).

The relationship between intimacy, liberty and the reproduction of Eurocentric bourgeois power depended on the simultaneous reproduction of subjugation. That subjugation was organized through race and Empire. In European colonial contexts, the symbolic over-sexualization of Indian, Chinese and black women in accordance with the sexual desires of Western men had the function of establishing the intimate household order of white female domesticity (Nandy 1983: 32; Stoler 2010: 8). Nineteenth-century US images of the intimate-feminine as chaste, modest and meek operated similarly through the over-sexualization of black femininity as the constitutive threat to white intimate order (Carby 1987: 27). As the white woman symbolized the intimacy of the family home with its connotations of civility, domesticity and property, the Indian, Chinese and black woman symbolized its negative opposite, often through her appeal to the white man's baser instincts.

The intimacies of colonized people were also structured by racial and colonial orders. Servants and slaves were less availed of time and private quarters in which to ruminate on their inner selves, and familial, romantic and kinship attachments were structured differently. Indeed, the intimate itself was often a site of brutality and control, as accounts of sexual violence reveal.[6] As Lowe notes, just as there was a colonial division of humanity there was also a colonial division of intimacy, 'which charts the historically differentiated access to the domains of liberal personhood from interiority and individual will, to the possession of property and domesticity' (Lowe 2015: 18).

But intimacy nonetheless persisted among colonized and racially subjugated people, albeit in an alternative historical relationship to the development of white bourgeois modernity. To deny that is to

---

[6] The mind, the body, sexuality and sexual organs, friends, family and homes are sites of racial and colonial intimate governance (Beauvoir and Halimi 1962; Stoler 2010; Weizman 2007).

limit human intimacy to one set of normative relations privileged by racial dominance, and simultaneously thereby to erase the possibility of intimacy from the subjects of racial and colonial violence, which is also to remove the essence of humanity. As authors like Frederick Douglass and Primo Levi make clear, the fact that intimacy persists in systems of crushing racial violence is testament to its enduring human quality (Douglass 2001; Levi 1988).

However, in the social sciences and humanities the tendency has been less to understand the racial and colonial contingencies of intimacy than it has been to shoehorn intimacy, and much else, into liberal and Eurocentric dichotomies of private and public, femininity and masculinity, interiority and exteriority, and emotion and rationality, and use that as the basis for normative discussions on well-functioning democracy (Habermas 1989).

These much-repeated approaches have ushered in a phalanx of analytical errors. These include the notions that: the public is not intimate; public relations are not concerned with interiority; too much intimacy (privacy) causes a lack of social connection; and too little social intimacy produces alienation (Dean 2005; Putnam 2000; Turkle 2011). In this confusion, the solution to these modern malaises is sometimes also intimacy (Giddens 1992: 3). The tendency there is to counter a patriarchal version of the public that is not intimate by championing the equally patriarchal designation of feminine virtue as intimate and private – something long since cautioned against (Beauvoir 2009 [1972]: 16).

Lauren Berlant bridges the false distinction between the intimate female private and the rational male public by engaging with the intimacies of late modern media cultures. Working with the concept of the 'intimate public', Berlant explores the ways in which affect is consumed publicly (Berlant 1998; 2008; 2011; see also Linke 2011). This helpfully disrupts the valorization of intimacy as positive or negative and contingently opens latitude to think of intimacy as both affirming and damaging; as coexistent with patriarchy and bourgeois capitalism, not an inner-worldly cure to it. But Berlant is nonetheless hampered by figuring intimacy as a benchmark for well-functioning secular society, and her earnestness on these terms misses many of intimacy's more productive alternative registers.

This book's approach to intimacy is different. Through developing a relational approach to intimacy, it avoids the analytic dead ends of binary analysis. Intimacy's qualities depend on its relations not on a priori categories.

In psychological works intimacy is said to occur when an individual has self-knowledge (intimate and internal) and is able to share this knowledge with another such that it may be reciprocated (Fisher and Stricker 1982: xi). Intimacy is, then, the product of relation, of an interaction between two people, as Wilner comments:

> The ability of each intimate to be a full presence with the other bespeaks a reciprocity between them so fundamental that it need only be hinted at by the barest of looks, smiles, or intonations of voice in order to be immediately and unmistakenly grasped. (Wilner 1982: 24)

Intimacy is not a question of public or private but at essence a product of *relation*, of a reciprocity so fundamental it can be *intimated* in the barest of gestures. Expanding from two people to a collective, this reciprocity can colour the entire atmosphere of a social situation, such that it frames the interaction itself rather than being merely the product of it. It can also persist in the absence of the gestures that produce it, such that it precedes those acts. Here it becomes not only produced but also productive: an 'active process of shared engagement rather than simply a feeling or state' (Sexton and Sexton 1982: 7).[7]

As much as intimacy is produced and productive in collective social situations, it is also maintained through objects. In this book we are considering sound objects. In her essay 'Where Might I Find You?', Gail Lewis explores this dimension of intimacy. She considers how music (Jamaican popular songs) and their associated technologies (78, 45 and 33 rpm records, and audio tapes) were the intimate sound objects through which she experienced her relationship with her father within a changing racialized and diasporic north-west London context that was black, white, Jamaican and English. For Lewis, those sounds and sound technologies are intimate because of the relationships they sustained with her father and with herself, because of the way in which she related to those materials and because of the ways in which personal feelings were carried through them (Lewis 2012: 138).

---

[7] What Raymond Williams might refer to as a structure of feeling (Williams 1977).

In this way, Lewis is alluding to how intimacy is conveyed through sound and sound technologies, but also how intimacy is figured in post-colonial dynamics. That is to say, how intimate relations are intrinsic to the working out of being and belonging in social and cultural histories characterized by displacement, poesis and movement (errantry) (Glissant 1997: 160).

Intimacy's foundation is, then, relational. But to deepen our appreciation of intimacy as a concept we must also elaborate its other particular qualities. The bulk of scholarly works on intimacy sidestep or presume intimacy's make-up rather than elaborating on it. They tend to employ intimacy in place of affect or as a popular term to collect dynamics of private life. This limits the analytic potential of the concept.

Developing a thicker conceptualization of intimacy requires expanding on its defining qualities. The defining qualities of intimacy are: presence, knowledge, depth and wholeness.

For Lewis, we see how the intimate *presence* of sound and sound technologies was in tension with the uncertain presence of her father.

> Maybe even at around age six I was still using the music to hold me together, that perhaps it was precisely the musical sound that I had taken into myself as a binding force that I could rely on in the uncertainty of your presence. (Lewis 2012: 143)

Discussions on intimacy's *presence* are further documented in literature on intimacy's non-verbal dimensions; the powerful spells cast by the gaze, the face, the kiss and the touch (Lorde 1984: 329; Marar 2012; Prager 1995). For Levinas, the intimate presence of the Other's face is the basis through which you understand yourself, not on your own terms, but in relation (Levinas 1989: 75–87). Indeed, the intimacy of that presence can be so profound that on its basis you may die for the Other (Levinas 1998: 2–3).[8] Intimacy's presence is not only held in the face, the gaze, the kiss and the touch but also

---

[8]Whereas in the existentialist works of Fanon and Sartre the gaze over-determines the Other, ultimately rupturing the internal ego (in Fanon this is the racialized Other and the white gaze), in Levinas there is no foundational ego prior to relation to be overdetermined in this way (Fanon 1986: xxix; Sartre 1969). The presence of the face-to-face is pre-ego and pre-cognitive, such that you are not *understanding* the other on the basis of the self, which Levinas sees as an act of violence, domination, possession and property, set on the terms of the knower (Levinas 1998: 8–9).

in music and sound. Levinas was additionally interested in 'whether rhythm's impersonal gait – fascinating, magic, is not art's substitute for sociality, [and] the face' (Levinas 1998: 9). It follows that the intimate presences of the rave or dancehall might be powerful enough to dispel difference and eliminate personal space between people.

Intimacy concerns *knowledge*. Indeed, the colloquial use of intimacy in relation to knowledge is well established – that one might have intimate knowledge of a person or of a situation, for example. Knowledge (and particularly knowledge of the Other) can be violent – concerned with owning-as-property the being of another. That version of knowledge is sometimes referred to as 'information' (Benjamin and Arendt 1968: 91).[9] However, intimate knowledge also exceeds information. 'Beyond knowledge and its hold on being, [does not] a more urgent form ... emerge, that of wisdom' (Levinas 1989: 78). Wisdom is the intimate knowledge arising from the fullness of shared life, of knowing touched by a plurality of hands. We have already encountered this version of intimacy in the discussions above. Johnson's 'bass history' is wisdom, not information. The 'moving hurting black story' does not fit neatly within modern data streams.

Intimacy pertains to *wholeness* because intimacy entails apprehending the Other as whole, whether they are present or not. As existential philosopher Marcel writes:

> 'Even if I cannot see you, if I cannot touch you, I feel that you are with me; it would be a denial of you not to be assured of this'. *With* me: note the metaphysical value of this word, so rarely recognised by philosophers, which corresponds neither to the relationship of inherence or immanence, nor to a relationship of exteriority. It is of the essence of genuine *coesse* ... that is to say, of genuine intimacy. (Marcel 1948: 25)

Intimacy here is the state of experiencing another's wholeness without consuming it, without making it property, without

---

[9] "The value of information does not survive the moment in which it was new. It lives only at that moment; it has to surrender to it completely and explain itself to it without losing any time. A story is different. It does not expend itself. It preserves and concentrates its strength and is capable of releasing it even after a long time' (Benjamin and Arendt 1968: 90).

designating it Other and without exhausting it in time. Wholeness relates to both knowledge and presence in that intimacy 'is immediate *whole* knowledge and presence that awaits being evoked' (Wilner 1982: 24, my emphasis). Wholeness can be realized in a society of beings present to one another – 'in an *intimate* society' – and it can also be realized through their relationship to sound and sound technology (Levinas 1998: 16). Intimacy's *depth* takes temporality into consideration. Intimacy's depth is 'great time' (Bakhtin 1986) – the flows of prior moments that constitute the now. Depth might be experienced as expansive (as Johnson's 'bass history') but it can also be immediate and intense (the jungle rave). Depth might then be the longue durée of an alternative modern flow, as it might also be experienced through the rhythms set by of property regimes of the now.

Sonic intimacy

The subject of sonic intimacy receives various renderings in scholarly work. One of the most authoritative is Tia DeNora's *Music in Everyday Life*. In that book, DeNora explores the intimate dimensions of music (codified sound) to understand the role that it plays in romantic relationships. She explores how music is culturally constructed along with ideas of love, gender and sexuality and how, for heterosexual couples, their performance of love and sexuality is intertwined with the symbolism that music has on these terms (DeNora 1999; 2000). Certain songs are symbolic of certain constructions of romance, and the playing of them helps us make sense of ourselves as gendered beings in sexual relationships.

As part of the wider textual trend in cultural theory, DeNora's work helps us to understand why a man or woman might choose a certain song in a certain moment – because it symbolizes their gendered understanding of love and physical romance. But, at the same time, it makes it difficult to understand the ways in which sonic intimacy might exceed the symbolic. It makes it hard to know what intimacy feels like in those moments, or, indeed, how part of that feeling might concern our intimate relation with the music (or music technology) itself.

Grossberg's work on music and affect sets off on a different foot. For Grossberg it is music that conditions the social, providing parameters and possibilities for the rhythms of everyday life (Grossberg 1991: 363–4). Grossberg argues that music 'calls people emotionally' as 'the most powerful affective agency in human life', which 'almost independently of our intentions' produces and orchestrates our moods (Grossberg 1991: 363). For him, verification of this is musician Carlos Mejia Godoy's prediction that the Contras would lose to the Sandinistas in the Nicaraguan Revolution because they had no singers. He was right, but whereas DeNora over-privileges the textuality of sound, Grossberg over-privileges its affectivity, making it equally hard to understand how symbolic aspects of music relate to the Nicaraguan Revolution.

In a discussion of dance music, Gilbert addresses this impasse by making a case for sound and music as signifying and affective.

The very existence of musics which exist primarily to be danced to suggests that any attempt to talk about music-in-culture must have recourse to an understanding of music as effective at the corporeal level, and not merely as an exercise in signification. Indeed, even the most cerebral music of the concert tradition must be understood as working affectively, at least in part: unless music is merely read as an unheard, the strings which vibrate the air which vibrates our skin, membranes and bones communicates a force which is not the same as the cognisable message encoded in the pitch intervals and rhythm. (Gilbert 2004)

The above quote highlights how the affective, felt or sensate dimensions of music exist with, and cannot be fully accounted for in, discursive, textual and semiotic analyses. So just as we must hold sound and signification together, so too must we hold affect's place within and through them. A consideration of affectivity is, then, not a denial of the textual or the signifying. Sound cultures are sonic/affective and signifying/intimate. If these connections are not made, the place of signification itself becomes confused. Although the non-linear, sometimes pre-cognitive character of intimacy cannot be fully encompassed in the narratives of signs, it cannot also be denied that sign systems in sound cultures contribute to intimacy; that coded meanings have affects not contained in language (Massumi 1995: 86).

Following Deleuze's revision of affect theory, this position is now better established in sound scholarship (Deleuze 1990).[10] Affect is not solely found in pre-cognitive dimensions of sound but also in signification. We find affect, for example, in the narrative of the music, the darkness of the dancehall, or the sight of the reggae speaker stack and its becoming relation with the reader/viewer (registered as excitement/apprehension), just as we find it in literature, language and the spectacle (Deleuze and Guattari 2004: 84–5 – see also Henry 2006; Massumi 1995; 2002; Smith 1997: xxiii). If the dancehall is viewed as only sonic and affective and sound is afforded a unique relation to affect, it becomes difficult to understand what is at stake as the visual becomes central to black diasporic sound culture, that is to say as YouTube exerts its gravitational pull. We might be fooled into thinking that, with the advent of YouTube alone, we need switch on our textual lens. That is a mistake that misses the longer and more telling fluctuations of signification, affect, sound and light in black diasporic sound cultures.

Unpicking the hold of semiotics and textuality on modernity through an analysis of sound is not then to discount the affect of the semiotic, it is not to assume that the textual and the visual are one and the same, or that there is a neat delineation between the visual and the sonic. Rather, it is to understand that the sonic has visual imaginaries just as the visual sings. Images play through your head when you listen to pirate radio just as sounds can play through a painting viewed on a wall, and both are affective. The visual has textual, phonic and affective properties as sound has phonic, textual and affective properties, and the two are in relation. This allows for the ways in which the YouTube screen sings and it allows for the ways in which the singing screen alters the sonics of a grime song.

Some of these debates are played out in scholarship on soul and post-soul music in the US and UK. That scholarship is concerned with the transformation overtime of black radical aesthetics, intimacy and politics. Leaning on a broader cultural studies framing, soul is conceptualized as a 'structure of feeling' located in black popular

---

[10]'A sign, according to Spinoza, can have several meanings, but it is always an effect. An effect is first of all the trace of one body upon another, the state of a body insofar as it suffers the action of another body. It is an affectio, for example, the effect of the sun on our body, which "indicates" the nature of the affected body and merely "envelops" the nature of the affecting body' (Deleuze 1997: 138–9).

culture at the end of the US civil rights era marked by the death of Martin Luther King Jr. and associated with a new consciousness and culture (Iton 2008). However, while social struggle marks the social context, it is music and more particularly the politics and intimacy of soul music to which the concept by the same name refers.

As registered in Aretha Franklin's 'Think', released in 1968, soul is both signifying and felt (Iton 2008). It is social and relational in that it extends from a period of social struggle against racism and white supremacy. This informs its wholeness, which permeates and connects multiple social-sonic nodes. Soul is '"with", it is on the road ... in the company of those that feel the same way, "feel with them, seize the vibration of their soul and their body as they pass"' (Deleuze 1987: 62).[11] Its shouts and screams have sonic histories that exceed the codification of music, that are irreducible to 'verbal meaning or conventional musical form'. The screams and shouts of Aretha, James Brown and others open onto a history of joy and suffering extending to the world (Moten 2003: 6).[12]

---

[11] Soul is not trapped in the private domain, 'a more or less chaotic welter of happenings which we do not enact but suffer (*pathein*) and which in cases of great intensity may overwhelm us as pain and pleasure does' (Arendt 1978: 72). Arendt derives this assessment from her own binary assumptions of public political action and private life, a distinction she sees broken down and internalized through a socially conformist/totalitarian modernity in a manner that is incapacitating for political action (Arendt 1958).

[12] This is sometimes explained through the account of Aunt Hester's scream. Aunt Hester is slave woman who appears in Frederick Douglass's *Narrative* (Douglass 2001). She is family member of Douglass's who, on account of her freedom-seeking actions, is whipped. Douglass hears the whipping and the scream, and that forms the basis of a double movement in his work between the scream (sound) from Aunt Hester and the slave song (music). 'While on their way, they would make the dense old woods, for miles around, reverberate with their wild songs, revealing at once the highest joy and the deepest sadness' (Douglass 2001: 21). Moten works with the same double movement to draw attention to a form of phonic materiality irreducible to 'verbal meaning or conventional musical form' that plays through ruptures of the slave triangle in ways that are temporally conditioned by the concrete moments of the violence, wrought by racial capitalism and patriarchy but not held/contained by it (Moten 2003: 6). As he then addresses the 'sonic' of *sonic intimacy*, he also opens the 'intimate' because in that double movement of soul a private ego is not overwhelmed or political action denied (as Hannah Arendt would have it), but respectively a joy and suffering extended powerfully and politically to black people, and (for Du Bois) the world. This reverberates in the screams of James Brown, Abbey Lincoln and in the broader fullness of black diasporic culture. With regard to Lincoln, 'Where shriek turns speech turns song – remote from the impossible comfort of origin – lies the trace of our descent' (Moten 2003: 22).

Scholarship on post-soul is concerned with the ways in which soul transforms through the music machines of late modern black diasporic sound culture. For Kodwo Eshun, the relationship of humans to modern technology is qualitatively changed by the introduction of drum machines and sequencers. In post-soul music such as jungle, humans are then said to no longer move through machines but are moved by them. Writing on the Roland TR808 rhythm composer, whose pirated software versions became a staple of jungle production, Eshun discusses the ways in which the technology opened up a 'new threshold, the programming of posthuman rhythmatics ... of programming humanly impossible time', which seize and rewire 'the sensorium in a kinaesthetic of shockcuts and stutters'; a machine-located intensification and pleasure driving a generation (Eshun 1998: 79, 186–7). The implication here is that the human is no longer expressed through technology but is merely its physical extension.

This book's understanding of the sonic intimacy of soul and post-soul music does not follow Eshun down the post-human rabbit hole. Rather, it maintains that, as drum machines and sequencers are introduced, humanity extends through music technology. Jungle dancers are not seized and rewired through its sound technologies but are contingent with and productive of them – as jungle's technology is also productive of the human. Nor is the jungle producer split from their technology. They are not empty vessels wired into 808s (McKittrick and Weheliye 2017: 31). Rather, jungle's drum sequencers become the expressive means through which life's pains and pleasures play out in all its non-reductive complexity. Those machinic flows include joy and sadness, and might include violence and emancipation.

That brings us to our second reading of Ellison, and his essay 'Living with Music'. Ellison has been introduced above as a seminal thinker of the sonic and the technological. And with a gendered lens applied, we can also engage with his contribution to intimacy in ways that help clarify the approach taken in the proceeding chapters of this book.

'Living with Music' is an autobiographical account of a 1950s American apartment building Ellison shares with an opera singer (Ellison 1995). The opera singer occupies the dwelling above. The story begins with Ellison positioning himself as a mediocre trumpeter and in the traditions of jazz and jazz men. That builds into a dialogue between jazz and opera and also between Ellison and the opera singer, that acknowledges not just their cultural and racial differences but also the similarities of their institutions, practices and expressions. Through music the apartment building is elaborated as a microcosm of the contradictions of American life; a location through which the neat dichotomies of race and culture (black and white, jazz and opera, unmodern and modern) can be productively called into question. One of his devices for this is technology, with his modern hi-fi and jazz recordings prominent in questioning the Eurocentric myth of modernity.

This evaluation of modernity and race through sound extends to intimacy and in particular the intimate *presence* of sound. This is achieved through his ensuing contest with the opera singer. He describes how the singer's voice, which travels from the floor above, gets 'beneath his skin'. He talks of her notes 'shifting through the floor and my ceiling, bouncing down the walls and ricocheting off the building in the rear' (Ellison 1995: 228–9). His response is channelled through his hi-fi, through volume and jazz, which melds into an American cacophony. In this way, he engages with the ways in which sound moves intimately through his body and through the building, tying together a personal, social and historical scenario, and by extension rendering the contradictions and porosities of race in everyday American life.

However, that account of sonic intimacy in his apartment building reveals more than an astute engagement with sound, intimacy and the contradictions of race in modern America. At the same time (although it is not Ellison's critical intention) it shows how patriarchy also plays through these relations. Although they are not the point of the essay, masculinity and violence are in fact central to it. The essay is as much a discussion of race as it is an engagement with mastery, control and domination. The contest with the opera singer is after all violent. Ellison weaponizes amplified music to reassert his dominance over their shared building, and over the opera singer herself. As he does so he demonstrates his mastery over technology; over the hi-fi, which has corollaries with the jazz men's mastery of their instruments. The male affinity with technology as modern vector of control and domination in Fordist (and post-Fordist) society is played out.

Here again we see how sonic intimacy is socially produced and culturally informed, and how the relationships of humans, technologies and sounds contain popular entanglements of conviviality and violence. Reading Ellison's 'Living with Music' in that way provides a bridge to understanding the sonic intimacies discussed in this book, and their conjoint commentaries on race, everyday life, and masculinity's 'intimate bond with technology' (Wajcman 1991: 137).

How the book proceeds

This chapter has been used to outline how sonic intimacy is understood in this book, and to signal what is at stake in sonic intimacy for black diasporic sound cultures. To address these concerns in more detail, each of the following chapters explores a different sound culture and conjuncture – reggae sound systems, jungle pirate radio and grime YouTube music videos. Each establishes the sound culture through its technological, social and sonic relations, and explores its sonic intimacy with regard to presence, knowledge, wholeness and depth. That discussion is structured such that the chapters need not be read in any particular order. If you're more interested in jungle or pirate radio than the other subject matter, you can begin there. However, to understand the overall argument on transformation all the chapters need to be read. After all, the book is less a guide to each discrete sound culture than it is an argument about the movements between them.[13] That argument, along with a polemic defence of the approach taken in this book vis-à-vis contemporary race scholarship, is collected in the Conclusion.

Moving across the chapters is then an evaluation of what is in play, as sonic intimacies transform through late modernity. The book addresses the shifts from the depths of dub to the intensity of

---

[13]This is also to say that the book should not be read as a comprehensive guide to the development of black diasporic music culture in the UK. A fuller picture of that dimension of musical history can be found elsewhere (Melville 2020). Neither is it a guide to reggae, jungle or grime, although it draws on substantial materials from each. Those wanting more detail on those music cultures are encouraged to follow the references in each of the corresponding chapters. Nor is this a history book. While the book develops historical sources, its purpose is not to recover a deep history. Above all else, then, this is a book on cultural theory, and should be approached as such.

the YouTube screen, and the presence of the pirate radio phone-in to the hyper-locality and deterritorialization of grime YouTube music videos. It assesses the extent to which alternative cultural politics are intertwined in racial capitalism and where they exist beyond them or below them. The intention is to provide an assessment of the state of sonic intimacy in late modernity, and also to present a detailed argument on why we should give a damn about it.

In more detail, Chapter 2, 'Vibe and the Sound System' explores the sonic intimacy of reggae and dub sound systems. The book makes a case for understanding the sound system as sonic and intimate. That entails establishing the sound system in terms of its techno, social and sonic relationality – through understanding the ways in which the sound system is relational on social, technological and sonic terms, but also with regard to the ways that each of these facets interrelate. Having foregrounded the relational dynamics of sonic intimacy, the chapter elaborates the conjuncture in which the sonic intimacy of the sound system must be evaluated. This conjuncture is marked by the post-war crisis of capitalism and heightened institutional and street-level racism. It is through this context that the alternative black cultural politics of the sound system are engaged. From here, the chapter explores the sonic intimacy of the sound system. That discussion is broken down under a number of headings, some of which correspond to other chapters in the book and some of which are peculiar to the sound system. The first dimension of sonic intimacy discussed in this chapter is the kinship relation that comprises the sound system. That leads into a discussion on craft and the sound system's intimate labour and material relations. The narrative moves from there to the dancehall, exploring notions of intimate proximity in terms of sound, space, bodies and social location. This is complemented through debating the intimate knowledges contained in those dances, and indeed in the dub itself. Those evaluations of proximate and intimate knowledge provide the basis for an appraisal of the development of urban multiculture in that period. The wholeness of the sound system is discussed in terms of its vibe. Depth is explored through the notion of 'dubwise'.

Chapter 3, 'Hype and Jungle Pirate Radio', explores the sonic intimacy of the jungle pirate radio in 1990s London. Whilst there is a growing literature on UK sound systems, jungle pirate radio, the metronome of mid-1990s London, is sparsely documented.

This chapter recovers that story through jungle pirate radio transmissions. To do that it develops pirate radio's particular history in terms of its techno-social-sonic relations. Those are located in the appropriate social conjuncture. That conjuncture is characterized by the consolidation of neo-liberal capitalism under John Major's Conservative government, by the ongoing development of urban multiculture and by high levels of institutional and street level racism. Starting with a discussion of kinship and community, Chapter 3 moves to consider the intimate presence of jungle pirate radio as it existed between raves, bedrooms, cars and broadcast infrastructures. The chapter proceeds to explore the alternative knowledges of jungle and, with regard to labour, the implication of a shift from craft to DIY. From here, there is a sustained discussion of hype, which is to say jungle's condition of wholeness, and of depth, which in jungle is a question of immediacy and impermanence.

Chapter 4, 'Grime and YouTube Music Videos', is the final sound culture discussed in the book. This chapter addresses the post-2008 moment in which YouTube music videos become the gravitation centre of grime sound culture in Britain. This requires a review of the longer history of video and black diasporic sound culture, including an exposition of earlier VHS video culture in reggae and jungle. The social conjuncture of YouTube music videos is established through an elaboration of the crisis of neo-liberalism, a crisis that has been slow burning throughout the post-war period. Those discussions provide the framing for a detailed evaluation of the sonic intimacy of grime YouTube music videos. With regard to intimate presence, the simultaneous deterritorialized and hyperlocal presence of grime YouTube music videos is addressed. The analysis of intimate labour relations reflects back on previous discussions on craft to explore what the closed-circuit DIY and prosumption models of YouTube imply for black diasporic sound culture. From there consideration is given to grime sound as it plays through pre-2008 sound ecologies – pirate radio and independent record stores. Attention is paid to four of the grime's key registers: grime, coldness, agonism and intensity. Those, while not exhaustive, are argued to combine into the wholeness of grime sound culture, captured in the term 'grime' itself. That discussion continues into an elaboration of grime's treble (and MC) sonics and the lo-fi ethics and aesthetics of

YouTube music videos. The chapter ends with an exploration of the effect of the YouTube screen and networked platform on the sonic intimacies of grime, with attention paid to the legibility of race on the screen and the attendant affective shift occasioned by the interface itself.

Each chapter contains an evaluation of what is at stake in the transformation of sonic intimacy for black diasporic sound culture, and for society more broadly. The book's Conclusion consolidates that discussion. It does so first by substantiating the political merits of a relational analysis of sound culture vis-à-vis the standard modes of left critique in the field of race and post-colonial studies. That polemic clarifies the importance of opening and sustaining questions about the existence and potential of alternative cultural politics to dominant racial capitalism. The second half of the Conclusion addresses what is at stake in the shifts in sonic intimacy explored by the book. Specifically, it considers how relation, craft, great time and wisdom, finitude and mutuality and wholeness, have been and might continue to be, without guarantees, important forms of sound culture, and useful points of engagement for those concerned with how sound cultures sustain alternative registers for human life to that re-scripted in dominant racial capitalism.

2

The reggae sound system and vibe

Introduction

In 1981 *NME* magazine undertook the somewhat impossible task of itemizing UK reggae sound systems. Their list included a hundred London-based sounds, and more than thirty elsewhere across the country, only a proportion of those active at the time (NME 1981a). That simple act of enumeration spoke of the social significance of the reggae sound system at that moment. A list that missed off many smaller sounds nonetheless showed that up and down the UK sound system culture was at its apogee.

Today we can reflect on the significance of this moment. Although, back then, there was some disagreement about the UK sound system's influence on social and political life, now, whether it be through sound production, sound technology, sound culture, music or in terms of the racial politics of Britain, it is undeniable that the sound system has profoundly influenced how we listen, produce and indeed *are*. The sound system has fundamentally altered the way that Britain relates to itself. Reverberating through popular culture, underground music, anti-racist activism and street protest, the sound system produced a demand that exceeded the racist state at that moment. Central to this was bass, the expansive sensibilities of dub and the deep co-presence of the dancehall. The contention of this chapter is that these reverberations concern *sonic intimacy*.

Consider the following quote from Channel One sound system's Mikey Dread:

> I can tell you about sound system from now 'til tomorrow morning put you have to feel sound system to know what it's about. (Mikey Dread interviewed in Folke and Weslien 2008)

Here, as in much other commentary on sound systems, the sound system cannot be spoken. Verbal description falls short. Mikey Dread instead places emphasis on what you *feel*. To know a sound system is to feel it.

While revellers should heed Mikey Dread's advice, filling themselves with the unspeakable qualities of the sound system, the task for writers is somewhat different. We still require words. Mikey Dread then proposes a conundrum for the sound system scholar – how to write about something that surpasses language? How to analyse in text the sound system's feeling?

One of the most convincing approaches to resolving this problem has been to explore the tactility of bass (Jasen 2016). Here, Mickey Dread is taken to be referring to the feeling of the bass on your skin. When he says, 'you have to feel the sound system to know what it's about', he is taken to be alluding to the touch of low frequency sound waves on the human body. I agree with that analysis, but surely feeling does not stop there? We can say that because, while the sound system is evidently a question of the pressure on the flesh, it is also much more. After all, the feeling of the sound only makes sense within a broader feeling for the sound culture, a feeling for its knowledge, for the music and for the people present with you in the room. In other words, the feeling of the sound system refers not only to bass but to *sonic intimacy*.

This chapter explores the sonic intimacies of reggae and dub sound systems by reworking scholarly commentaries for threads abandoned or only partly unpicked. It similarly engages with documentaries, films and interviews. This chapter is roughly located in the 1970s and 1980s, a period that coincides with the movement between Jamaican and black British renderings of sound system culture, and the development of alternative cultural politics by a generation of black and working-class people growing up in the UK. Here we are shifting from rocksteady to reggae and dub sound systems, and from smaller sounds like Metrobeat[1] to the larger sounds of Jah Shaka, Sufferer Hi Fi, Fatman, Frontline and Earthquake. This is the high point of UK sound system culture.[2] These huge bass-resonant hi-fis are the gravitational centres of a largely self-sustaining black British music culture comprised of record importers and store owners, [1] label owners, pressing sites, cutting houses and distributors (May 1977a; 1977b; 1977c). Across UK cities, weekday dances are culminating in weekend parties. Community halls in Seven Sisters Road and Tollington Park feed into town halls in Tottenham, Hornsey, Wandsworth, Brixton and Acton. The Apollo in Willesden, and All Nations in Hackney, Four Aces in Dalston, the Crypt in Deptford and the Rialto in Birmingham supplement pubs like the Bluesville in Wood Green, many student unions and countless house, basement and youth-club dances (Gayle 1974; Sullivan 2014: 61).[3]

Relation

This chapter explores how reggae sound systems were comprised, experienced and lived in that moment and what that entailed for the alternative cultural politics of 1970s and 1980s Britain. It tells this story through sonic intimacy: through the sound system's intimate social and material relations, through its intimate knowledges, through its proximities, through the fullness of its vibe and through its deep sensibilities. But in order to consider those matters we first need to engage with how the reggae and dub sound systems were dynamically comprised technologically, socially and sonically. We need to acknowledge the basis of sonic intimacy in the sound system's techno, social and sonic relations.

---

[1]This period of sound systems, including the stories of Duke Vin and Count Suckle, has received some attention (Bradley 2000; Melville 2020; Salewicz 2012; Vin 2017).
[2]This is a predominantly masculine sound culture. Lovers' Rock was a more mixed gender reggae space and has been discussed by Bovell and Back 2017; Palmer 2011.
[3]Although much of the discussion in this chapter is based on London sound systems, there was a sound system culture in black Caribbean communities across the UK (Campion 2017; Huxtable 2014; Jones 1988; Riley 2014; Veal 2007).

In the same 1981 special edition of *NME*, Sir Coxsone Outernational[4] was asked to produce a technical specification of their sound system (NME 1981b).

Sir Coxsone Outernational Sound System – Technical Specification

Amplifiers: five pieces of 600 watts (4 × 600 bass valve amps, 2 × 600 treble transistor amps) – depending on venue i.e. Brixton Town Hall determines three pieces of valve and one piece of transistor.
Preamp with built in equaliser to cover weight, treble, midrange.
HH Echo Unit.
Special percussion box.
Speakers: on average 19–20 bass speakers.
Many different horns in treble section and small speakers.
Several thousand yards of cable.
Seven-ton truck and transit van.

Each line of the specification, which includes a seven-ton truck and transit van, denotes a component of the sound system. As the reader scans down the list, the sound system emerges from a collection of separate units. We piece together its power, its frequency, its weight and size and something about the quality of its sound, too. But this list is also an illusion. The technical specification is an act of accountancy that enables the control and auditing of the sound system as much as it facilitates a cultural appreciation of its make-up.

A cultural understanding of sound system technology is not a technical specification. Rather, a cultural approach requires that we place its parts in dynamic relation, understanding the sound system more as a system than as the pieces of a puzzle. The technology of the sound system is inescapably together. Speaker cable is not only PVC-insulated multi-stranded copper wire but the conduit for transmitting power from the amplifiers to the speakers. The speaker

---

[4]Named after the legendary Jamaican Studio One producer, Coxsone Dodd, and owned by eponymously named British Jamaican sound system pioneer Lloyd Coxsone.

is defined by its relation to other components. And, as such it is not an alarm, a siren or a personal address system but a bass, mid-range or treble unit.

This is not where relationality ends, because the social also relates to the technological. That is to say, people and their practices are inflected in the wires, speakers and so on. The operator who runs the 'set', the selector who plays the tunes and the deejay (the MC in post-rave nomenclature) responsible for the lyricism give definition to the technology, controlling the microphone or running and configuring the set. The technology gives meaning to their roles, too. The microphone defines the deejay and the record defines the selector.

The point I am developing, and which will run through this book, is that the sonic intimacies of the sound system can only be known in relation. If we do not foreground that we end up with a technical specification of culture, one oriented to particular professional or indeed scholarly systems of control, which often say more about the mastery of sound culture itself than they do of its dynamics.

The sound system's relationality does not end with the individual's reciprocal relationship with technology either, because these individual social roles are co-dependent. The sound system is not generally a solo pursuit but that of a collective, of the crew. As there is a reciprocal relation between microphone and deejay, there is between sound system and its crew. The crew gives the technology a particular status, and the power and quality of the sound system gives meaning to them. In his book *What the Deejay Said*, Lez Henry discusses these techno-social relations, explaining how the sound system shared a reputation with the deejay. In his case, this was a reputation of upliftment, conscious lyrics and black empowerment. It was not the deejay alone who carried this reputation, but the whole sound system, such that a system would be evaluated as a techno-social collective (Henry 2006: 151).

All of these relations (including the box boys who carried and moved the equipment, the prospective deejays and selectors and, for larger systems, the electronics or speaker-building specialists) are cohered through sound. It is in fact sound and not technology that announces the sound system. It is sound that calls across the dance floor. As sound calls the crew and technology into being, it also verifies the collected roles therein. If you are the operator, your business is sound quality, power and reliability. If the sound is poor, then you and your technology are so appraised. When Henry mentions the reputation of his sound system for upliftment, he is addressing sound. He is identifying lyrical discourse (which is sonic) and also implying music (also sonic). This sound can further be divided into frequency ranges that correspond with the drums, instrumentals, lyricism and bassline.

But without the dancehall, this means little. When active, the sound waves emitted from the speakers relate the technology and crew to the audience. The sound system presupposes the audience who – far from being passive recipients of music – are co-producers. The sound system is predicated on dialogue with its audience. The sound system in this sense is a techno-social configuration dependent on the liveness of its social relations for its definition. The technology of the sound system attains its meaning (its reputation, in Henry's terms) through its relation to the audience, who verify its stance through dance and voice; through somatic and sonic symbols. The skank of the dance floor and the call of 'rewind!' are essential information for the sound system crew. When a sound system doesn't play out, where it resides un-strung, it may still have a crew, but it ceases in fact to be a sound system. Locked in a warehouse, garage or basement, it loses its social life, its vitality, becoming again an assortment of prone technology, with potential.

Crisis

Having established the relational basis for the sound system, we must foreground the conjuncture in which it is active. The cultural and political dimensions of the reggae sound system can only be evaluated when placed in their social and economic context. For the sound system, that conjuncture was shaped by the post-war crisis of capitalism in which racism and nationalist assertion featured heavily.

The structural crisis of capitalism defined the post-war period in which sound systems played, shaping the 1970s and 1980s as it had the two decades previous. Global economic shifts had led to a decline in Britain's manufacturing base, reduced technological competitiveness and a weakened global trade position. Those shifts found a resolution in Conservative Prime Minister Margaret Thatcher's free market revolution, and contingently in racist and nationalist claims often levelled at black people and black culture, including sound system culture.

After a choppy first term, the Falklands War paved the way for Thatcher's unexpected victory in the 1983 General Election, providing the political opportunity[5] for rapid deindustrialization, the contraction of the state and the rude implementation of neoliberal market economics. Drastic cuts in public spending, soaring inflation and unemployment ensued, impinging heavily on black people, the majority of whom were working-class in this period. Fifty per cent of the general youth population was unemployed, with black youth worst affected.

The economic resolution to the crisis is accompanied by a racist one. Prior to Thatcher, the post-war state had expanded significantly into economic and social spheres, meaning that economic strains were also experienced as political ones. The state had made itself responsible for large tracts of the economy and society and it was thereby also culpable for its shortcomings. Those political strains pulled at the legitimacy of the government, leading to an erosion of social democratic principles. From the tenure of Labour's James Callaghan administration (1976–9) onwards, this erosion was increasingly filled with populist politics characterized by an authoritarian appeal to law and order. In place of a social democratic model of consensus, thin appeals to the people became the mode through which the government sought to rule (Hall et al. 1978: 214).

Those populist appeals to law and order drew from the deep well of British racism, often siphoning anti-black sentiment to its ends. The ideologies of Otherness – through which Britain had legitimated its rule in the colonies and through which the Edwardian and Victorian upper and middle classes had designated the working-class – provided a reservoir for this populist vitriol. Political leaders

---

[5]Thatcher reformed her 1979 cabinet, removing the older more consensual guard broadly loyal to former Prime Minister Edward Heath, and installed her own evangelists.

like Margaret Thatcher[6] and Enoch Powell[7] propagated variants of white supremacist, nationalist and exclusionary politics to consolidate political power. In so doing, they bolstered the far right and entrenched the spectre of black criminality.

Finding outlets in popular culture, notably through Eric Clapton and David Bowie,[8] this cacophony of nativism and anti-black racism resonated through Britain, and was thickened by colonial actions, namely the British state's involvement in Northern Ireland and then in the Falklands (James and Valluvan 2018). Throughout the 1970s and 1980s black people were discriminated against in private and council accommodation, subjected to over-policing, arbitrary arrest and detention, police violence, victimization for reporting crime, specious armed raids, violence and sexual humiliation in police stations. In education, black children routinely faced harassment and institutional racism (Institute of Race Relations 1987). There was the indiscriminate use of SUS laws (which saw black and Asian people stopped disproportionately by the police), culminating in the SWAMP 81 operation that saw 943 people in Brixton stopped under suspicion in four days. The police failed to properly investigate the actions of the far right in Southall, East Ham, Bradford, York and many other locations (Institute of Race Relations 1987). Their

---

[6]Thatcher's speech on 'swamping', televised for Granada TV's *World in Action* in 1978, bolstered support for National Front and its rejection of racial outsiders. But more importance still is placed on the influence of Enoch Powell who, prior to Thatcher, drew on the US white power politics of the late 1960s to amplify discourses of anti-black racism to substantial political effect.

[7]Enoch Powell was a Conservative MP for Wolverhampton South West and in his 1968 'River of Blood' speech he warned prophetically of the consequences of black migration to the UK. Developing a 1964 General Election speech on evils of the 'colour question', he drew on US white supremacist discourses around 'black ghettos' and the 1967 'race riots' to portray black working-class people as apolitical, anti-social and detached from mainstream British society. In the immediate aftermath of the speech, racist attacks took place involving a slashing incident on a black family by fourteen white youths shouting 'Powell' and 'Why don't you go back to your own country?' See for discussion Hirsch 2018.

[8]From 1975 to 1976, David Bowie, one of the world's largest pop stars, expressed sympathy for Nazi ideology, making a Nazi salute (which he argued was a wave) outside Victoria Station. Bowie later apologized for these errors, but we should not discount the significant impact that such a self-professed dalliance had for black and Asian people in the UK. Indeed, along with Eric Clapton's white supremacist rant, it was sufficient to spur the formation of Rock Against Racism in 1978 – which included such performers as The Clash, Tommy Robinson Band, X-Ray Specs and Steel Pulse.

failure to investigate the deaths of black children in the New Cross fire became a particularly potent rallying point.[9] To say the police were both racist and belligerent in this period is an understatement,[10] and indeed this knowledge became popular fare for reggae's more politicized sound systems.

The absence of employment for young people, and the danger they faced on the streets at the hands of the police and racist gangs, also placed greater importance on the affordable leisure provided by the sound systems. With unofficial race bars operating on the doors of many clubs, sound systems were some of the only sites black young people had to congregate around (Jones 1988: 127). As focal points of black and working-class culture, these leisure sites were contingently seen as a threat to law and order; associated with a dangerous and culturally corrosive black culture. As early as 1957 when shebeens came into the white public consciousness, there were calls for their clampdown. The vandalizing by the police of Duke Vin's sound system in the 1950s heralded a litany of such mistreatment (Vin 2017).

Throughout the 1970s, there were routine police raids on parties, cultural events and youth clubs attended by young black people. The list of these aggressions is impossible to enumerate here, but to give a sense, early in the decade:

> The walls of the Church Hall at Faraday Road, Notting Hill, are covered with posters listing the battle honours of the last year ... the Metro Youth Club (sixteen arrested), Harlesden (eight), Acton Park (ten), Peckham Rye Fair (over fifty), Lisson Grove; and so the list reads on. (Hartley 1971)

---

[9] In the New Cross fire thirteen children were burned to death at a party. In the aftermath, police consistently and over years frustrated a proper criminal investigation through manipulation of the media, coercion of witnesses and the misassignation of blame to friends of the dead. These deep injustices resulted in the work of the New Cross Massacre Action Committee which organized a protest of 15,000–20,000 black people, and worked to hold the police to account and find justice. John La Rose was chair of that committee (La Rose 1984: 3).

[10] These injustices gave rise to a series of protests and uprisings in Brixton, London 1981, 1985; Chapeltown, Leeds 1975 and 1981; Handsworth, Birmingham 1981, 1985; Notting Hill, London 1976; Southall 1979; St Pauls, Bristol 1980; Tottenham 1987; Toxteth, Liverpool 1981.

We can add to these, the spurious raid of a black pub in Villa Cross, Handsworth and the police driving a car at high speed through a group of black people at Leeds Chapeltown Bonfire Day celebration (a special event for the black community in Leeds) (Institute of Race Relations 1987: 6; Plummer 1972).

A particular focus of police violence was the Mangrove Restaurant in Notting Hill. On one occasion during the 1977 Notting Hill Carnival[11] police entered the premises with indiscriminate force. Testimonies of victims from the time record how the police broke limbs, terrified children, administered savage beatings, frustrated emergency medical attention from ambulance crews and forced some to jump from windows 4 to 5 metres off the ground. Notably, they also targeted the sound system. Basil from the Black Punch Sound System was playing at the time.

> The first I heard was somebody outside shouting that the police are outside chasing people. Somebody else shouted to close the door. Next thing I know was a charge. The window next to the door was smashed by truncheons beating on it and then they came in. They started beating everybody and telling us to get out, but we couldn't because the door was cordoned off. I was thrown across the room. Police smashed my amplifier on to the floor and tossed my records all about. There was a couple with the child near to me and I protested to the police to let them get out. Apart from my amplifier, at least 45 singles were smashed and five LPs which you can't buy in this country at all. It's just nonsensical to me that they come into a place in that way, beating people and smashing property and making no arrests. (Race Today Collective 1977: 27)

---

[11]Corollaries to this include the 1970 Mangrove 9 incident in which nine members of the black intellectual and activist community were arrested for protesting against the continual raids of the restaurant on false drug and criminal possession charges. Those arrested included former *Race and Class* editor Darcus Howe. In 1974, the police had attacked the Carib Club in Cricklewood, arresting twelve people, among them the Sufferer Hi-Fi operator Dennis Bovell on a charge of incitement to riots, for which he received a sentence of three years in prison, before being acquitted after six months (Johnson 1981: 10–11). This event is immortalized in the final scene of *Babylon*, for which Bovell wrote the soundtrack (Rosso 1981). For more details on Notting Hill Carnival, please see Gutzmore 1993.

Kin and community

In *Bass Culture*, Lloyd Bradley writes that the sound system was much more than a wall of speaker boxes; it was quite literally 'the community's heartbeat' (Bradley 2000: 4). While Bradley's image of the sound system as the community's heartbeat is perhaps not the right metaphor for the relationality I have been outlining above – entrapped as it is in a hermetic biological image of organs in the body – it is a useful point from which to start to consider the sound system as intimate, in the context detailed above.

The film *Babylon* illustrates this well. *Babylon* follows the daily tribulations of a young man called Blue, played by the Aswad singer Brinsley Forde. Blue is a deejay with a sound system and the plot follows him and the sound system as they prepare for a dance. That provides the narrative arc to explore cultural production, racism and inter-racial friendship in late 1970s London. The same sound system that gets shut off by the police in the final dancehall scene has also entertained at a wedding in a local community hall. That is to say, the reggae sound system is portrayed as at once intimate to family events and blues dances. It has a privileged role in ceremonial accompaniment and in politicized soundings.

The privileged relation of the sound system to black British life is underlined by its versioning in the family home, and indeed in the most sacred of home spaces – the front room. Here, it took the form of the radiogram, and in particular the Blaupunkt Blue Spot. The Blue Spot was not a sound system. It was a radiogram. But the radiogram was nonetheless informed by its larger cousin; a domestic version of it. In the 1950s, radiograms were common among English middle-class families. Many featured drinks cabinets, making them entertainment hubs – music, drinking and socializing wrapped into one. While companies like Grundig and Garrard made popular models, Caribbean migrants often bought the Blue Spot because it was louder and had better bass response. It better suited listening practices informed by the music and sound system culture of the Caribbean.

Mikey D and Jah Baddis, sound system personnel from Coventry, both describe the central importance of this sound technology to the black family home, and the recurring preference for this particular model: 'We had a Blue Spot gram which most West Indian people did at the time' (Mikey D interviewed in Smith and

Porter 2010); '[the Blue Spot] was played on a Sunday afternoon' (Jah Baddis interviewed in Smith and Porter 2010).

The British Caribbean front room, in which the Blue Spot sat, was a hybrid of the Caribbean and English Edwardian-Victorian front room. The Caribbean front room was already a hybrid of colonial Victoriana and African diasporic aesthetics (Hall 2009). In Britain, its particular arrangement marked status and belonging in the UK, as addressed by British Vincentian artist Michael McMillan:

> You weren't allowed in this room unless there were guests, but it's Sunday, Jim Reeves's 'The Distant Drums' is blaring out of the 'Bluespot' radiogram and big people are chatting news from back home. Mum is drinking Babycham. I pass her the plastic pineapple ice bucket and listen quiet as a lamb, my skin sticking to the plastic covering the PVC imitation leather settee. I smell rice and peas and the paraffin heater competing with the air freshener and polish from the drink's cabinet filled with glasses that are never used. I touch the painted glass fish and plastic flowers on fanciful sugar-starched crochet on a gold-rimmed fake marble coffee table. A blue-eyed Jesus looks down at me from The Last Supper on floral wallpaper saying 'cleanliness is next to godliness'. Sunshine beams through pressed lace curtains onto the colourful patterned carpet. The front room looked so good, that it didn't matter how poor we were, we were decent people. Growing up, this room echoed the proverb, 'By the sweat of your brow, you will eat bread' and haunted me like nails scratching a blackboard. I know now that it is part of who I am and unconsciously I recreate a similar shrine in my own home today. (McMillan 2005)

The image McMillan paints is of a sensory space, filled with visual symbolism, deeply personal, deeply intimate to who his family were and who he is. Post-war American popular music is opening onto black British sounds, and in the midst is the Blue Spot radiogram – an image of a sound system focusing and channelling black family life.

Craft

The Blue Spot found its way out of these spaces, too, as younger family members repurposed its speakers and amplifier for their burgeoning sound systems. These practices illustrate the sound system's condition as DIY technology (a status shared by all technologies discussed in this book), and they also start to identify it as craft (a status not shared by all sound cultures in this book).

DIY approaches to sound technology have a long history in black diasporic sound cultures and, indeed, more broadly in sectors of society limited by wealth but not by creativity. They are part of the story of the jumper as football goalpost, and the plank as cricket bat. They are the repurposing of the ubiquitous and affordable turntable as hip-hop instrument. And they are the practices behind the initial emergence of sound systems in downtown Jamaica – who played recorded music for the masses, in part because uptown live bands were too expensive. But not all of these can be considered craft, and in craft there is something particular at stake for our discussion of sonic intimacy.

In the sound system, the DIY process is also a craft process, and the craft process entails intimacy between the object and the craftsperson – in this case the soundperson. The documentary *Sound Business* follows Sir Coxsone Outernational Sound System showing the process of cutting, gluing and screwing together new speaker cabinets. Alighting on the craftsperson, the interviewer enquires about the practice. Having spent days building the boxes, the craftsperson thus explains to the camera his labour, his relationship with the sound technology and with the sound itself. 'When you hear that section [of speakers] play', he says, 'the bass that going to come out of it, you are going to feel it. It is a crucial part. Hearing the sound play' (Dineen and Caidan 1981). Much more than tautology, the box builder is explaining to the camera how his craft becomes expressed in sound. The moment the sound system plays, the moment you feel the bass, is already prefigured in his intimate relation with the sound technology; with the act of gluing wood together. His expressive labour relation with the sound technology foreshadows the creative dynamics of the dance. Craft itself is sonic expression.

Craft entails skill and creativity. Sound systems, and particularly home-built and maintained sound systems, require the learning of specialist knowledge. 'A speaker cabinet have to be designed to suit the particular speaker you are going to put in it', says cabinet designer Mr Herron, 'you can't just make box and put a speaker in it. It don't normally work' (Mr Herron interviewed in Smith and

Porter 2010). Likewise, the sound system is tuned to the resonant space of the dancehall, with the deeper frequencies coming out of the back of the boxes allied to the architecture of an individual building. 'In the early days we worked hard on the frequency', explains Jah Shaka, 'not only power, to get the right frequency. Some bass is very loud, but some is Hz – rumbling but you don't hear – very low frequency. We try to get that' (Jah Shaka 2014).

This is not necessarily the learning of a science, although it may well be that, too – electronics, for example – but more often an art – a qualitative, interpretive and creative understanding of how the sound system will play.

All sound systems look for a combination of quality and power. When sound reproduction gets upwards of 3,000 watts, distortion or a blurry sound is a problem – frequently unsolved even in larger nightclub systems. With sound systems, this problem is compounded by mobility. Unlike a club PA, the sound system is only set up a few hours before a dance. Nonetheless, even in these testing acoustic contexts, power, quality and clarity are prized in maintaining the vibe of the dance, and each sound system fine tunes those attributes to achieve their characteristic sound.

Anybody who has developed a skilled relationship with an object that they deem expressive will understand the intimate significance that object, or collection of objects, hold. In an expressive art object, one that is prefigured in dialogue with the audience, this is magnified such that the sound system becomes a collectivized 'extension of personality' (Jah Observer interviewed in Gaonach 2014).

It is important to note, then, because it is not often said, that for sound system personnel, when the sound plays, the audience is not only listening to their music, but to their craft, which is an intimate relationship between craftsperson and object. To put it another way, if you have spent weeks building electrical components for a sound system, or even maybe just the frequency crossovers, when the sound plays and the capacitors hold out, your pleasure in the sound, in the dance and in the expression is also pleasure in the crossovers themselves; your intimate relationship with the sound system is at once with the crossovers and the revellers assembled.

These skills are learned through apprenticeship.[12] Junior personnel learn not through the canned immediacy of information manuals, but through what is shown, be it circuitry, pre-amp or cabinetry design. Here, there is an assumption that one cannot simply acquire the skills of the sound system, they are not available off the shelf, because craft is slow, learned and patient. It is wisdom.[13]

These practices sit askew to the dominant rhythms of capitalism, providing alternative potential for imagination and pleasure. The intimate relationship between craftsperson and craft object is out of step with capitalist production, enabling alternative forms of expression to materialize. The expressive content of the craft process and its characteristic attention to detail then forsake dominant rationales for production. In this quote from Count Spinner, an early sound system pioneer from Coventry, we see how the craft process operated on a tangential plane to dominant capitalist profiteering:

> I came to England in 1961 and I had a job at a hospital. The first week's wages was five pounds and I went and bought a little gramophone and I went and found a record shop selling blue beat records … And then I got my second wage and I took it to a timber shop and went and bought some board … After I finish making this sound system box, the House of Joy, a massive wardrobe, and the thing was we couldn't get it out of the room. It was too big.
> (Count Spinner interviewed in Smith and Porter 2010)

The fact that most people involved in sound systems have some version of this story is testament to the alternative relation of sound system craft to dominant capitalist timeframes. It is indicative of a form of obsession and patience, misplaced otherwise in capitalist money-making which, in industrial form, maximizes passivity and speed. This sentiment is well put by socialist craftsperson William Morris:

---

[12] For a discussion of apprenticeship in Jamaican sound systems, see Henriques 2011: 94–9.

[13] 'Craftsmen take pride most in skills that mature. This is why simple imitation is not a sustaining satisfaction; the skill has to evolve. The slowness of craft time serves as a source of satisfaction; practice beds in, making the skill one's own. Slow craft time also enables the work of reflection and imagination – which the push for quick results cannot. Mature means long; one takes lasting ownership of the skill' (Sennett 2009: 295).

> The creation of surplus value being the one aim of the employers of labour, they cannot for a moment trouble themselves as to whether the work which creates the surplus value is pleasurable to the worker or not. In fact, in order to get the greatest amount possible of surplus value out of the work ... it is absolutely necessary that it should be done under such conditions as make ... a mere burden which nobody would endure unless upon compulsion. (Morris cited in Thompson 1976: 646–7)

The speaker that was too big to fit in or too big to get out of a residential building does not fit this model. We might laugh at the calamity, but it is a serious point that this obsessive level of attention, an intimate investment in the craft object, disrupts the scripts of capitalist life. The sound system is an object of enthusiasm that exists against the grain of financial stratification; at odds with the imagination of a lucrative life (although they may also have desired this).

Presence and wisdom

The intimate presence of the sound system further disrupts the scripts of capitalist alienation. 'The performers are no different, in terms of socioeconomic background, from the people in the crowd; the music is performed and consumed by equals', explains sociologist Les Back.

> The MCs are by no means superstars, aloof and mythical ... The physical and social reality common to the audience and the performers alike makes the music relevant and accessible. The dance provides a powerful unifying context for the sharing and the celebration of collective experiences – the *power* of the music lies in its collective nature. (Back 1988: 149–51)

In the sound system, shared experiences are then powerfully cohered through sound and through horizontal ludic proximities. These are at odds with the individualism and privatization of Thatcher's Britain. The sound system keeps people together in the context of racial and classed alienation by providing a site in which daily experience and common social and historical circumstance can be collected. As the sound system collects and connects people, it verifies them through sonic and lyrical practices, reinforcing the sense of co-presence. As Back notes, sound system deejays would validate the mutuality of the space by relaying 'the ordinary humour, triumphs, tragedy, and despair of everyday concerns' back to the audience (Back 1988: 149–51).

These forms of presence were so central to the sound system that they were carefully replicated in the symbolism of the dance. Lloyd Coxsone explains:

> 'Without having those boxes and amplifiers the way we have it, it wouldn't create an atmosphere', he explains. 'You understand? It does create an atmosphere with the wires stringing over and the many boxes. You have to keep the sound in those boxes as it is because that represent sound'. (Lloyd Coxsone interviewed in Dineen and Caidan 1981)

Coxsone is drawing attention the ways in which speakers, wires and people are arranged reinforces co-presence. The wires connecting the speakers to the amplifiers are hung from the ceiling, creating a web above the dance floor, providing a canopy for the dancers below as the speakers form a corral.

Circulating through those spaces were intimate knowledges pertaining to the prevailing social context and to a wider diasporic oeuvre through which black people in Britain understood their relation to the Caribbean, to colonialism and to racism. 'Dread' of the 1970s was channelled through symbolism, sounds and interpretive atmospheres. The rudie and rebel met black radical, African anti-colonialism and US civil rights politics (Jones 1988: 52). Deejays used patios to vocalize Rastafari and the teaching of Marcus Garvey. Bob Marley, Burning Spear, Peter Tosh, Angela Davis and Martin Luther King merged with news of anti-colonial struggles in Angola, and a British analysis of racist SUS laws and working-class struggle (Arte Vost n.d.; Howe 1973: 18). That intimate knowledge reverse-traced the Atlantic region, establishing new marks on its surface. 'As the sound system wires are strung up and the lights go down, dancers are transported anywhere in the diaspora without altering the quality of their pleasures' (Gilroy 2002 [1987]: 284).

These intimate knowledges were by inference not aloof or disconnected but your 'daily livity', 'the topic of the day'. Blacker Dread explains,

> It was what was going on innit. That was the topic of the day so everyone would be involved in that because that was your daily livity. It wasn't nothing new. You weren't going there to learn something you didn't know. It was just enhancing and making people aware of the SUS laws and things like that. The deejays used to tell you about them 'man you go to the road and police stop you and them a gonna plant you, or them a gonna do this to you or dat to you'. In the New Cross fire, we were there on those marches. 1986 playing Anti-Apartheid – 100,000 people in Clapham Common. We played at those things. Eventually we became members of the ANC and we played for the ANC at Notting Hill Carnival for 17 years non-stop, every year. We didn't play on the street we played on the back of a truck. But our thing was politics. What we were doing there was making sure that people were aware of what was going on. Standing outside Barclays Bank in Brixton and telling black people not to go into Barclays Bank because it was an Apartheid thing. We did all of that and it was a part and parcel of your daily living because it was what was going on around you. It was nothing different.
> (Blacker Dread interviewed in Bass Culture Research 2018)

In the dance, these intimate knowledges circulated as wisdom. Sociologist and reggae deejay, Lez Henry explains, the deejays possessed the ability 'to locate me within the narrative [and] through their particular skills as wordsmiths actually had me considering whether the thoughts were theirs or my own' (Henry 2006: 201). Through the sound system, ravers and sound system personnel then knew each other's minds, developing what Jah Shaka refers to as 'intuition' – the ability to know another's mind – and 'telepathy' – the ability to do so through diasporic space and time (Jah Shaka 2014).

These wisdoms were carried in the deejay's words, and in the phonics of the sound system. 'Bass history is a moving/is a hurting black story', Linton Kwesi Johnson reminds us, directing attention to the affinity of bass notes with the black diasporic experience (Johnson 1980). In do doing, Johnson simultaneously recalls the double bass in jazz, the Fender guitar in Jamaica and the recording techniques of Sylvan Morris in Studio One (Bradley 2000: 309–10; Veal 2007: 98)[14] as he also tells of the 'freighted, frightening, fraternal "lower frequencies" of democracy and beyond' (Callahan 1996: xxxviii).

But so, too, were these more immediate wisdoms. Blood Shanti from Aba Shanti Sound System comments:

> I was born in England, with inner city vibes. So many artists strive for a Jamaican sound, or an American sound, but I feel that is the wrong road for UK artists to take. We've been born here and we *must* create our own sound forms, and put our stamp on the world. Actually, when we create music we reflect and echo what we see all around us, from the point at which we wake up every morning and look around us: when I go outside, I see city smoke everywhere, traffic, darkness, no birds, no trees, no nature. I know that some Jamaican musicians see our music here as too hard and harsh, but we're in a concrete jungle, you overstand, and I can only express myself from the way I live. (Whitfield 2002)

It is instructive to note that here the reggae sound system starts to convey a story of the black experience *and* a story of Britain itself – one that extends onward into the jungle and grime sound cultures also discussed in this book. This is not to say that Britain is benignly adopting black cultural form through the Aba Shanti Sound System – after all, sound systems were still too wound up with danger in the white popular imaginary for that to be the case – but at the same time, an intimate sonic knowledge of daily life, racism and class politics played by the sound system is shaping the livity of the multi-ethnic city.

On the dance floor, black, white and Asian found themselves in the Other, temporarily suspending the racist and colonial organization of British life. The knowledges in circulation mapped across the categories of immigrant and native maintained for the ends of populist rule. The non-proprietary wisdoms of the black diaspora resonated with the experiences of class and racist

---

[14]Ellington popularized the upright string bass (the precursor of the electric bass), preferring it to the tuba for its more explosive qualities. He also introduced paired bases, experimented with an array of bass musicians and introduced the 'crowding the mic' technique to get the most explosive sound from the instrument (Pekar 2010).

marginalization of working-class and Asian young people in Britain. The extent to which they did varied from place to place. Dances in Coventry, Birmingham (Balsall Heath) and different neighbourhoods in north-west, south and south-east London had different dynamics in this regard (Back 1994; Bradley 2000: 254; Coxsone 2018; Jones 1988: 120). The sound system was no crucible. A smelted amalgam of the city did not pour onto the streets as the dances called time. In the context of heightened racism, questions persisted around how black cultural symbols could be used and adopted by white people, what was acceptable and what was not, just as white prejudice did not evaporate day over night. But this did not discount shared pleasure of these spaces or the multicultural openings and porosities that they occasioned.

Vibe/ration

> Sometimes at a dance 'I don't know what the problem is, could be the place, or could be the sound you are playing with, could be anything but you find you don't feel right, the *vibe's* not right'. (Bikey from Sir Coxsone interviewed in Dineen and Caidan 1981; my emphasis)

For sound system personnel, the vibe was often presented as a mysterious and unknowable condition, present or absent in ways that surpassed explanation. But whatever it was, vibe was desired. Everyone in sound systems knows that a good vibe means a good dance.

The unintelligibility of the vibe is best accounted for by its condition as sonic intimacy, and through its particular quality of wholeness. The vibe is the unfathomable sense of a sound system's fullness; the feel of technology, sound and human participants in mutual relation. In *Endless Pressure*, Ken Pryce provides a good approximation of the sound system's vibe in a contextual passage on a Bristol blues dance (Pryce 1986).[15] The blues or shebeen, of the

---

[15]This passage sits within a somewhat essentialist criminological text on black deviance (Alexander 1996).

kind Pryce is discussing, was a dance, generally a private home with one or more sound systems rigged up. They were popular in British cities throughout the 1960s and 1970s. The Bristolian occasion described by Pryce was a reggae blues attended by people from the local British Caribbean community, with smaller numbers of white men and women present.

Pryce writes how in a 'good blues' there were usually 100 to 200 people tightly crammed into a single-floor flat. The music's volume made conversation impossible, so most communication was done through the body. This took the form of dance moves, but also the routine negotiation of the crowded space:

> We ring the bell and are let in, only to become part of a dense, teeming, sweaty mass of humanity. A bluish opaque cloud of tobacco smoke and vapour floats above our heads. Perspiration and body odours mingle with the sweet, herbal smell of ganja smoke and the pungent smell of beer: a crippling acridity pervades the whole atmosphere. Inside we remain motionless in the crowded passage, but trying all the same to push our way in. Blacks and whites are trying to escape to the nearest room where the dancing is taking place, but that room is just as packed. It is impossible, we cannot move backward or forward ... I pull myself from the incoming crowd in the passage and escape to one of the rooms where the music is playing, equally crowded ... I looked around me. The two rooms packed with people opened into each other. In one of the rooms, there was only one soundbox or speaker. Here the atmosphere was lighter and the light brighter. There was more movement. The people were dancing rhythmically to the music. Pairs of people dance apart, thrusting their hands forward and backward as they do the reggae. (Pryce 1986: 100-1)

Pryce captures the vibe of the blues outlining its human, sonic and technological relations – the proximate bodies, the intimate space, the speakers, the sound and the air. People pushing past each other, dancing together, variances in volume, rhythm and technology together constitute the atmosphere.

Lez Henry expands noting how the dynamic relations of the vibe are also identified through visual symbols and sound. Recalling a Ghettotone and Frontline sound system clash, Henry describes the excitement of hauling Ghettotone's quad speaker box (4 × 18 inch woofers) into the venue, anticipating the sound and feeling of bass, only to be confronted with Frontline's seven quad boxes which were to overpower the smaller system. The mere *sight* of these boxes, says Henry, filled him with anticipation, let alone the 'experience of their earth-shaking qualities' (Henry 2006: 170).

The sights of speaker stacks, moving bodies and webs of wires in the half-light are then important components of the sound system's vibe, but they are also secondary to sound. In a Channel One session at Brixton Jamm, Mikey Dread is visible, faintly illuminated by a lamp that also lights the turntable. A minor spectacle, people turn to face him, the record, Garvey, H.I.M. (Haile Selassie) and the Jamaican tricolor, although also turning to engage the people in their midst. But as sound moves through the space, visually over-stimulated minds relax and the bass, toms and lyricism take over. Like the tiny red LED load indicator atop the enormous speaker stack, the visual becomes so minimal, so unadorned, that it is only a faint pressure amid the weight of the air and the vibration on the body.

The vibe of the sound system is predominantly sonic and the most significant component of that is bass. Bass 'causes a vibe … energy flows through the bass', says Shaka (Arte Vost n.d.). Although it is not plausible to separate the bass from the drums – the bass provides the roll, the drums insert the groove – it is the bass that is always in ascendance. The drums cannot exist without the bass. They are sustained by it. But the bass has a different relation to the drums. The bass is the bedrock, enhanced by the drums, patterning their rhythms and relentless without them. Russ Disciples from Boom-shacka-lacka explains:

> Reggae is a bass dominant music. The melody is in its bass line … the bass line is there as an underpinning … So you see our sound systems, our big speakers and massive amps and we push them things hard to make people feel it. (Russ Disciples interviewed in Folke and Weslien 2008)

Larger sounds might run twenty bass bins, sometimes in double or quad configurations producing 10,000s watts of unidirectional low end. This unidirectional effect is maximized by the configuration of the speaker stacks set up at the front and back of the dance floor in either quad or triangular formation.¹⁶ This is distinct from European concert consumption where sound is louder at the front than at the back. This sound system is not socially and sonically elevated. In fact, it is very different from listening to any other form of sound reproduction – headphones, home stereo, computer or mobile phone speaker, etc.

The waveform delivered by the bass bins is so slow that it's not fully picked up by the ears, but by the body. It is intimate with the body, moving within it, shaking your chest, your nose hairs and your internal organs. 'It turns questions of musical sound to matters of extra-musical vibration and extra-cochlear perception', crossing sensory thresholds (Jasen 2016: 3). Here, vibration *is* 'vibe' (Henriques 2008: 221). Lover's Rock singer, Sylvia Tella, recalls, 'I would be sitting in front of a speaker and you see their bodies going drvvvvvvvv and I wanted to experience that so I would feel the bass line go brvvv' (Marre 2011).

The bass properties then charge your body, connecting to the sound system and the people in your midst; compelling you in particular ways. As the *NME* wrote on Jah Shaka:

> It seemed that when the other sounds had done with their boasting and toasting, there would come a discreet hiss from the corner, and Shaka would mutter a title, or more often an invocation to Jah RasTafari, and the old-style heavy bakelite-style head of his arm would lower to the vinyl. Then it might seem that the walls were tumbling down around your ears. Then it might seem that your body had never felt those rhythms to impel and overwhelm, you'd find your feet flashing like sparklers. (NME 1981c)[17]

As your body is moved, so is your mind. Connecting the sonic intimacy of vibe and vibration to presence and wisdom, Steve Goodman, aka dubstep producer Kode 9, writes,

---

[16] Quad is two at the front and two at the back. Triangular is one at the back and two at the front.
[17] As with vibration more broadly, this full bodily experience can of course be translated in other ways; some people find it pleasurable (Trower 2012: 126–49), some people find it grounding, others find it deeply relaxing. I have a good friend who was on more than one occasion found asleep, curled up at the foot of a bass bin at the end of the night.

For many artists, musicians, dancers, and listeners, vibratory immersions provides the most conductive environment for movement of the body and movement of thought. (Goodman 2009: 79)

And, as with the rest of the sound system, this was not a lonely flight but one that was relational because those same vibratory encounters also worked to 'unravel self-certainty ... recast our sensed surroundings ... and rethink what we can do and how we operate' (Jasen 2016: 3–4).

Dub

The time and the space of the dub contribute depth to the sound system's vibe. Errol Thompson, King Tubby and others would use the B-sides of their 45 recordings for the art of dub, versioning the A-side tracks with emphasis on the drums and bass with the 'chordal instruments only occasionally filtering through' (Veal 2007: 57). By the early 1970s dub was a genre in its own right. Lee Scratch Perry's *Blackboard Jungle Dub* (Dub 1973), *King Tubby Meets the Upsetter at the Grass Roots of Dub* (King Tubby and The Upsetter 2005), and Sir Coxsone's *King of the Dub Rock 1* and *2* are notable albums here (Sir Coxsone Sound 1975; 1982).

Dub is the avant-garde work of sonic deconstruction, the tools for which are the mixing desk, reverb, echo, phasing and tape editing. As LKJ explains,

> Here, the recording engineer, through skilful manipulation of the controls, the use of echo, reverb, phasing and other sound effects in the mixing of the rhythm tracks, is able to lend the music an added rhythmic and illusory affect, making it particularly suited to dancing. (Johnson 1981: 10)

In the dance a second level of deconstruction takes place as the selector drops the bass in and out, adding reverb and echo to further distort time and space. Jah Shaka is unsurpassed in this aspect of the art. From the early 1970s he has presided over a four-decade exploration of distortion, weight, expansion and space through his sound system.[18]

> Sometimes [Shaka] plays the vocal section straight, then he rides the rhythm until it disintegrates, you hurtle through the instruments like a dance of swop-your-partners, now whirling to the hi-hat, or fist-fighting with the bass. (NME 1981c)

The dub in studio and dancehall guises is concerned with the creation of space. The emphasis is on subtraction, on pulling away at the edges of the music rather than building layers towards a crescendo (Simon Ratcliff from Basement Jaxx interviewed in Natal 2008). Through the creation of space, the musical centre of gravity shifts deep into the roots, moving away from simple melody and obvious rhythm. 'This sense of spaciousness, [tugs] away at the edges of the musical structure ... evoking a presence that hangs in the gaps between the notes' (Chambers 1985: 161). This space creates a sense of anticipation and feeling of depth.

> With the dubs, you're working with a rhythm that's hanging on the verge of collapse all the time. You're putting it to pieces, holding it together with delays and adding and spinning the rhythm, taking out ... one bar blurs into another or distorts into the end of the four-bar figure, and then you pull it back, just when you think it's gonna collapse. You soothe people by bringing back the bass when you've taken it out. There's more space in it than anything. (Adrian Sherwood cited in Veal 2007: 79)

This depth is compounded by dub's longue durée. The spatiality of dub's depth is twinned to illusions of time. Operating outside the normal formula for consumption and gratification in popular music – chorus, verses, crescendos and finales – the dub invites a long-term embrace, particularly in its live sound system formats. Live dub dances last for many hours; as Mikey Dread of Channel One notes, Channel One just doesn't play its best tunes quickly, 'we take you on a musical journey over 5, 6 hours' (Dread 2012).

---

[18]See for more detail on this moment (Partridge 2010: 98–152).

In the dance this leads to a feeling of losing time. The repetition of rewind (when the selector pulls the track back to the beginning) reinforces that. Deejays also work to intensify the dub, allowing it to slip into, before bringing it back from, collapse. Deejay Ras Kayleb's utterances match the slipping, before closing in on the trance of the dub, bringing you back to reality.

Alongside the deejay's work, the repetitive composition of the music and the extended playing of versions of the same track, the meditative embrace of the dub means that it is frequently difficult to recall how long you have been dancing for. In a diametrically opposed way to which industrial alienation is regulated by the hands of the clock, dub break the mould of capitalist time for alternative imaginings. This temporality is referred to as dub-wise.

The alternative imaginings of dub-wise are well discussed. For African Simba, the dub lifts you out of the rhythms of capitalist exploitation; distressing, winding down, lightening the load and getting by:

> A lot of people when they go to the roots dance, whether they Rasta or not, whether they go with the intention of getting rid of a burden or not, they say coming out of a dance, Simba I feel different. I feel like all my problem is gone. My shoulder feel light like I was carrying something and it is gone. (Afrikan Simba interviewed in Folke and Weslien 2008)

But there are other possibilities for the dub-wise, too, altering your imagined relation to time and space more fundamentally, putting you on a 'different level, a different planet', making you feel like a 'space man' or 'deep-sea diver' (Robbie Shakespeare cited in Veal 2007: 63), or indeed providing an opening to violence on the dance floor as the strictures and governance of the day-to-day are displaced, and the emasculation of daily life flows over (Back 1988: 150).

The sonic intimacies of the sound system

The sonic intimacies of the sound system are like no other discussed in this book. In the context of heightened racism and class marginalization their properties were such that they collected people, conveyed a demotic story and provided space for forms of expression and imagination that were at odds with the modes of oppression under which black people had lived and were living. This wasn't a marginal concern but one, from the home to the pleasure of the dance, that was central to black British life in the period; nor was it defined as a novelty supplementing the expansion of capital into ever-newer realms of commodification. Rather it was the centre of kinship, community and leisure life as it was also autonomous in terms of its processes of production and dialogue.

The sound system was a dynamic system of relation comprised of technology, sound and society, in which each facet of the culture was intimately present with the other. Socially, people with similar experiences of racism, class marginalization and daily life opened onto each other, as they found their shared stories played through the speakers and in the sound (both musical and phonic). Racist social structure impinged on these spaces, but so too were these presences sufficiently open so as to engender multi-ethnic relations on the dance floor, in the experience of sound and in society more broadly.

The capacity for forms of expression that moved outside capitalism rhythms was ensured by the sound systems' processes of labour and co-production. The crafting of the technology provided for expressive relations at odds with industrial alienation and Eurocentric consumption. They allowed for the maintenance of alternative traditions of pleasure, skill and attention, and a dance space for these to be collectively realized and co-produced.

Here alternative knowledge, intimate to those in the room, could breath and circulate. As wisdom was present through the process of crafting, so too did it move through sound and people in the dance. The people in the dance brought their experiences and histories into the room where they were collected under a web of wires, and there those experiences were intuited by deejays, and resonated in speaker boxes and through the black diasporic history of bass.

The bass told a collective story of black diasporic life, as it moved through revellers collecting them together with the speakers and the studios. Its vibrational properties engendered movement, pleasure and release as they also dulled the tyrannies of the visual, racial and capitalist, expanding mutual relation. The dub further deconstructed the sonic-sensory experience of Eurocentric capitalist time. If all that came together, the dance had a good vibe; a vibe so charged that it would flow through time.

3

Jungle pirate radio and hype

Introduction

> Goes out to all the massive locked in ... Shout out to all those who are getting cotch time to stick a tape in ... Sending the maximum moves out to the Guvnor. Hold tight Smurf. Goes out to the Ragga Twins, hold tight Flinty, hold tight Deeman ... Absolutely rolling 94.5 ... Goes out to all those cutting on the open road, time to take some element of control as we flex the vibe coming at ya. (DJ Brockie and MC Det 1996)

In this way, MC Det introduces DJ Brockie over the track '16 Track Ting' on Kool FM in the summer of 1996, recalling a moment in the early to mid-1990s when jungle pirate radio was the metronome of the city, setting a frenetic tempo at the end of a decade of Tory rule. The Criminal Justice Act had come in and the outdoor acid house raves around London were being closed down. The ravers, soul heads, b-boys and reggae crews, many of whom, black, white and Asian, were born in the late 1960s and 1970s, were mixing at clubs and warehouse parties in inner-city London. The black vernacular sound they compose, they call 'jungle'. As Rebel MC aka Congo Natty says, it was 'the people dem who called the music jungle'. At the same time, terrestrial pirate radio suffered another heavy crackdown; this time it did not recede, but exploded and stations proliferated like never before, or since, as thousands of young people in their living rooms and cars tuned in.

In the fast-changing landscape of the 1990s, jungle pirate radio was faster. That seemed permanent, although its maximum velocity was only really sustained for five years. Its struggles and pleasures were evidently traced from acid house and hardcore raves, but more closely still from the 1970s and 1980s reggae dub sound system. But jungle pirate radio's cultural politics were also different. These were the 'bastard children of Radio Alice': fierce but goal-less in their apolitical unity (Reynolds 2008: 234). Rapturous, kinetic, angry, warm and intense in equal measure, formed in the socially stratified and deeply unequal crucible of Tory Britain, their cultural codings were alternative but did not demand a new utopia. An affirmation of alternative existence but not a claim on the future as such, jungle pirate radio refracted, demeaned, repurposed and charged that moment's over-sped capitalism for London's minor key.

The sonic intimacy of jungle pirate radio was 'hype'. Part mobility, part liquidity and part speed, hype was fierce, fractured, fragile, live and illegal. From the setting up of transmission sites to the call and response, hype was sustained across the board. In 1995, Kool FM's Marley Marl shouts over the airwaves:

> 'Phone calls coming thick and fast' Footloose can you help me, please?' 'Stratford massive, Ilford massive, picking it up good over the east region. Short and sweet that's how we like it'. (Marley Marl and Footloose 1995: 8 mins)

Hype still contained a rebel stance – in the illegality of the endeavour and the black militancy of the music, in the patterns of dialogue sustained across the city, in the forms of multi-ethnic and spatial collectivity, and in the commitment to common good. But this was a people against ... I'm not sure what, culture. It had the all the angst, energy, escape and catharsis but no particular purpose. Guy Called Gerald, Congo Natty, MC Navigator, MC 5ive'O, MC Hermione and Kemet Crew's Mark X retained elements of black internationalist, anti-racist, civil rights and Rastafari idioms, but these did not predominate, or define the culture. After all, in the trauma and pleasure of John Major's Britain, jungle pirate radio's more routine demands were in-time or higher tempo even than contemporary capitalism's rhythms and desires.

Jungle pirate radio was the structure of feeling for a generation, never far from its headlines and media panics, but in the annals of cultural history that rearrangement of diasporic social and sonic forms is a mere murmur. Between the terminal burn of Tory Britain and the euphoria of early Blair, those contours of time are now footnotes to the heartbreaks and highs of Thatcher and New Labour.

Few cultural studies of jungle were published at the time,[1] and less still written about jungle pirate radio itself. Ethnographic approaches were frustrated by the radio pirates' secrecy, while media studies focused their preferred gaze on commercial, state, military and community broadcasters. For radio studies, black diasporic illegal music radio was not actively disregarded as such, just not heard.

To address this, the chapter engages with recordings of jungle pirate radio shows uploaded by fans to Internet sites. Discussion of these is supported by analysis of journalism, literature, documentaries and film footage from the period. Jungle pirate radio's online archive is uneven. Prominent stations such as Kool FM are better accounted for, whereas many hundreds of smaller stations from around London's peripheral market towns – the stations of High Wycombe, Uxbridge and Hertford, for example – are absent. In this way, we miss the young people who stuck an aerial up on a shed, or the side of a hill, to broadcast to the town centre massive.

History of jungle pirate radio

Appearing after the vibrations of the sound system and before the screen devotions of YouTube, jungle pirate radio is synchronous, not separate, from both. It shares the black vernacular music forms of reggae and grime, as it does their autonomous media ecologies and geographies of the city. Their modernities and mobilities are intertwined, too. In jungle pirate radio, the recorded music and transistor technologies of sound systems move through telephony, cars, motorways and young people's bedrooms.[2] Jungle pirate radio's history is, then, a modern one as it is also a black diasporic sonic one.

Popular attention to illegal music radio has focused on the UK sea pirates of the late 1950s and 1960s for whom the term 'pirate radio' was coined. The risqué but palatable heroics of stations such as Radio Caroline and Radio London are the stuff of exhibitions

---

[1] These studies included Christoloudou 2009; James 1997.
[2] Michael Bull addresses a different version of this modern sonic mobility story through the Walkman (Bull 2000).

and books recalling a rebellious and popular moment, ultimately absorbed back into a revamped BBC following the 1967 Marine Broadcasting Act.

The terrestrial pirates emerging in their wake are less well accounted for. Laser 558 played pop after Radio Caroline's demise. From the end of 1970s Invicta played soul, and by 1984 'the summer of pirate radio' Dread Broadcasting Corporation, JFM and Horizon were transmitting reggae, electro and soul all over London (Goddard 2011: 24; Hind and Mosco 1985). By 1988, thirty-one pirates were operating in London, including London Weekend Radio, Galaxy and Kiss FM. They played soul, reggae and other genres of black music (Wolton 2010). By 1989–90, there were over sixty pirates in London alone, including the hardcore, rave and acid house stations: Sunrise, Centre Force, Dance FM and Fantasy (Hebditch 2015; 2017).

Jungle pirate radio was mixed here. Many early jungle pirates were nurtured in sound systems and illegal reggae, rave, soul and funk radio. The Ragga Twins, Kool FM staples, were part of Unity Sound system. DJ Hype worked as a hip hop scratch DJ, made flyers and speaker boxes, listened to hardcore pirate stations and joined Heatwave sound system in 1984 to play ragga, rare groove, hip hop, soul and early house. While working at a warehouse in north-east London, he joined pirate radio WIBS in Tottenham before moving to Fantasy FM in Hackney. In 1989, PJ, Smiley and Hype ran ragga reggae sound system Private Party. Influenced by Soul II Soul's movements through north and East London, Shut Up and Dance (PJ and Smiley) released a mixture of house, hip hop and hardcore just as Paul Chambers (Ibiza Records) combined acid house with his dad's reggae, and Rage residents Fabio and Grooverider mixed soul with house, techno and hip hop (Belle-Fortune 2004; Evans 2014).[3]

---

[3] In addition, DJ Zinc is mixing pre-acid house electro, acid house and hip hop (Hyponik 2016). The Prodigy, from the suburbs east of London, are doing speeded up break beats, with techno and EQed basslines, and start to make appearances at Dalston's Labyrinth. MC GQ, whose brothers and uncle worked in sound systems, is developing hip hop lyricism on reggae sound systems taking forward what hardcore MCs Chalky White and Emerson Rat Pack had done previously (Belle-Fortune 2004: 150). MC Skibadee is listening to hip hop, rare groove, soul, rap and ragga and developing a jungle lyricism with the rhythm of KRS One, Public Enemy and Capelton.

By 1991 'jungle' was emerging. Defection FM introduces break beats over hardcore tracks (DJ Wicked et al. 1992). On Don FM, Ed Rush is playing 'hardcore' composed of digitized low-end bass, 160 bpm hip hop breaks and ragga MCing (Rush et al. 1993). In Harlesden, rude boys and crusties are getting down to something that's 'not hardcore' (Belle-Fortune 2004: 13). In the clubs and warehouses of Hackney and Tottenham, Early Rebel MC (aka Project X, aka Congo Natty) and Lennie Da Ice tracks are generating big crowd responses (Christoloudou 2009: 46). Jungle station Kool FM displaces East London hardcore pirate Centre Force as the most prominent London pirate radio station. Weekend Rush, Conflict FM, Cyndicut FM, Eruption FM, Flex FM, Format FM, Freedom FM, Rude FM and many more join them. By the mid-1990s, jungle is indisputably the sound of London, and by the late 1990s the sound of much of the East, South East, South West, Midlands and South Yorkshire, too.

Jungle pirate radio as relation

Jungle pirate radio was broadcast from studios in sheds, disused flats, squats and hired premises. Standard to most studios were two turntables, a mixer with crossfader (the influence of hip hop), an amplifier, speakers, headphones and a microphone. These studio technologies were connected to a transmitter. The transmitter was hardwired at the studio's location or, more commonly in this period, connected to a microwave emitter. The microwave emitter then sent the signal in a direct line of sight to a microwave receiver located on a different site. That protected the position of the station in the event that it was discovered by the authorities. The microwave receiver was connected to a solid-state transmitter wired to an antenna from which the radio signal was sent. The FM signal is reduced as it passes through solid objects, so these antennae were placed high up, often on the roofs of tower blocks.

On the reception end was a radio receiver. By the 1970s domestic radios were common in British homes. These radios were usually the size of a shoebox and contained one speaker that produced about 5–10w of sound. By the 1990s these sets were being replaced in young people's bedrooms by the hi-fis and portable stereos of Japanese technology companies Sharp, Sony, Akai, JVC, Technics

(Panasonic), Kenwood and Hitachi. Whereas the shoebox radio could adequately reproduce the human voice of BBC Radio 4, these sets claimed to reproduce X-bass, 3D bass, turbo-bass and the like at around 30–50w. Informed by sound systems, hip hop culture and jungle raves, they matched the changing listening practices of young people.[4]

> Big up to all the driving massive. (Marley Marl and Footloose 1995: 8 mins)

Car stereo receivers follow these trends. In the car stereo, the surround acoustical properties of the reggae sound system found affinity with a mobile micro-environment. Young men desiring power and potency fitted powerful but affordable Japanese-produced stereo systems into their otherwise inexpensive vehicles; accumulating amplifiers, headsets, crossovers, wiring, modified parcel shelves, mid and treble speakers and powerful bass units. These were secondary sites of jungle broadcast as urban and suburban high streets were filled with distorted bass lines and rattling car exteriors.[5]

> You could walk down the street and everyone was playing jungle from their cars. It blew up, it was a movement and you could feel it. (Brockie in Evans 2014: 11.50 mins)

The technologies of pirate radio stations were maintained and practised by collectives of people. Larger stations involved DJs and MCs, management (for schedules and revenue), and technicians (for radio installation and maintenance). Those roles were supplemented by a wider set of occupations. People in south London built transmitters and set frequencies, people in Finchley made the aerials and people at metal workshops fabricated boxes so transmitters could be locked on site.

---

[4]Their claims to bass, X-bass, 3D bass, turbo-bass and the like is testament to the sound system, but they have also moved away from the black Atlantic routes of reggae technologies. Driven by global commercial and consumer imperatives, jungle pirate radio is powered by Japanese sound technologies.
[5]See, for a related discussion, LaBelle 2010: 161.

Down with the Motorola. Yo! Down with the Motorola ... to the Nokia ... come harder // Down with the Motorola. Yo! Down with the Motorola, Mickey Finn and MC Shabba gonna make ya night come harder. (MC Shabba, performing at One Nation New Year's Eve party 1997, Cordell 1997b: 8 mins 15 secs)

Then there was 'the massive' 'locked in' using telephone technologies to phone in, text or leave missed calls to the stations. When read back by MCs, these facilitated dialogues that verified the pirate radio back to itself.[6]

These techno-social relations spiralled out of the sonics of the rave. Jungle pirate radio channelled the synthesized bass of the dance floor through a plurality of domestic radios that could not themselves get down to 30hz (despite their claims). Clubs like Roller Express in Edmonton and Telepathy in Dalston, raves like Exodus and United Dance, and numerous squat and warehouse parties would then send the tunes that generated the most hype back to Kool FM, Rude FM, Rush and the others, for wider public play. As those tracks played, jungle pirate radio championed the underground sharing its sounds and promoting its network of raves, clubs, record stores and artists. From 1991 to 1995, nearly all jungle tunes came into circulation in this way.

The power that these stations had in breaking new tunes was so intense. I would tape certain shows to hear new tunes and then head straight down the record store and spend all my money! (King Yoof quoted in JamieS23 2014).

Neo-liberal consolidation

This all took place in a period of neo-liberal consolidation. With the formal end of Empire and the Cold War, Western monetarism was now global-hegemonic, and programmes of privatization were busy dismantling the welfare state (Slobodian 2018). Traditional forms of family, work and collective life were being uprooted by de-industrialization and political economic reorganization.

---

[6]See for a detailed commentary of mobile phone use on pirate radio, Fuller 2005: 50–2.

Consumption and entrepreneurism, rather than production and mutuality, were the watch words of the time (Bauman 1998; Beck and Beck-Gernsheim 2001). Jungle pirate radio's apogee then coincided with a less secure, more precarious and more impermanent society, especially for its working-class protagonists less able to maximize themselves through spending power (Bauman 2000).

Between 1990 and 1993 housing repossessions accompanied housing benefit cuts and council housing was reduced through pro-market 'right to buy' schemes. Labour market reorganization saw low and semi-skilled employment deprioritized, disadvantaging working-class young people. Unemployment among the young working-class was high, particularly for black and Asian young people also facing structural racism (France 2007: 18; Mizen 2004: 67). In education, the introduction of free market competition and league tables, coupled with funding cuts, led to an increase in school exclusion among black boys.[7] Only one in five young people leaving school at minimum age (16) entered the labour market and only half of these started on government work schemes (Mizen 2004: 23–55).

This was also a time of heightened racism. In 1993 Derek Beackon won the BNPs first seat in East London's Isle of Dogs on a 'rights for whites' platform, targeting newly resettled Bangladeshis. In mainstream Conservative politics the Island Nationalism of Norman Tebbit MP was prominent.[8] These Little Englander positions fuelled street-level racism. In 1993, black schoolboy Stephen Lawrence was murdered by two white racists in south London.

The ongoing failure of the police to properly investigate these crimes was emblematic of its complicity, too. As with the decade before, urban policing in this period was known for its racist intimidation and sometimes racist deaths, too. The deaths in police custody of Joy Gardener and Roger Silvester were prominent rallying points (Mizen 2004: 131; Solomos 2003: 182).

---

[7]Permanent exclusions increased from 3,000 to over 100,000 between 1990 and 1997; the vast majority were boys, and the rates for black boys was five to eight times higher than those for whites (France 2007: 80).

[8]Tebbit's 1990 'Cricket Test' was fuelled by the fear that post-colonial (black and Asian) culture was supplanting English culture. The test of loyalty to English culture was, for Tebbit, who you support in a test match (Solomos 2003: 218–19).

Parallel to this, youth culture was labelled dangerous, debauched and out of control. Part of wider Tory moralizing on good virtue – that ended with Prime Minister John Major's affair with his colleague Edwina Currie – these moral panics were not triggered, as they might be, by the racist murder of Stephen Lawrence, but by the deaths of two more grievable victims – the white infant Jamie Bulger who was abducted and murdered by two older boys, and the Ecstasy-related death of white clubber Leah Betts (France 2007: 97, 136).

Bulger's death influenced the drafting of the 1994 Criminal Justice and Public Order Act (Criminal Justice Act), which doubled the maximum sentences for children in young offenders' institutions. Linked to concerns over dance music culture and drugs (and those of home counties' landowners) that same legislation cracked down on open-air raves, catalysing, somewhat ironically, jungle and the inner-city club scene.

In these panics, jungle was figured as dangerous and degenerate on black terms. In 1994, *Observer Life* magazine ran an article stating that jungle is a controversial and prominently black phenomena (Belle-Fortune 2004: 162). The police and local authorities link jungle pirate radio to black organized crime (BBC2 1994; London Tonight 1994). Even the name 'jungle' becomes associated with colonial notions of darkness, urban decay and 'ragamuffin house', informing a later shift to the whiter and more saleable moniker 'drum 'n' bass' (Congo Natty interviewed in Seely 1994: 11 mins; Ferrigno 2008: 7). Until Fabio and Grooverider appear on Kiss in 1994, and Radio One launches *One in the Jungle* in 1995, jungle is sidelined from broadcast.[9]

> We needed a platform to get the music out there and no one else wanted to play it. (Eastman, Kool FM interviewed in Mavros 2011: 1.58 mins)

Jungle pirate radio responded to this censorship by providing music for the masses. That commitment brought the pirates into contact with the authorities, their legislation and their regulatory powers. The 1990 Broadcasting Act expanded the

---

[9] *Mixmag* has a no jungle music policy (Christoloudou 2009: 109; James 1997: 58–9).

remit of The Telecommunications Act 1984, providing for two years' imprisonment for involvement in pirate radio, unlimited fines and the confiscation of studio equipment including the DJ's records. It also extended culpability under law to those indirectly supporting the station, and trading with it in any way. It justified this through the general panic around black and youth culture, and by specific concerns (propagated by Department of Trade and Industry) that pirate radio frequencies were leaking leak onto airport traffic control and emergency services bands – although the crystal locks of most transmitters and the shift of the emergency services to different frequencies made this unlikely (Trevor Dunn, Head of BBC Music Entertainment interviewed in Cordell 1997a: 1.40 mins; Wolton 2010). As Kool FM's Chef explains,

> It was bollocks … We were a gang, but we were a musical gang – nothing to do with causing trouble or selling drugs. (Chef quoted in Clifton 2015)

Kinship and community

Unlike the reggae sound system, jungle pirate radio was not intimate to the family. It was not present at weddings or found at the centre of the home. Indeed, as the next stage in black British youth culture, jungle was squarely defined against parental affiliations. Jungle pirate radio's intimate social relations of kin and community are better located in ties between friends, neighbours and youth club associates (Geenus interviewed in SBTV and All 4 2017: 8.50 mins).

These social intimacies were well reflected in the imagined communities of the stations themselves. 'Rinse' is the jungle vernacular for doing something well – 'Rinsing out a tune' – and in 1994 RINSE FM was set up as a jungle station by a group of East London friends (Geenus cited in SBTV and All 4 2017: 8.50 mins). RINSE FM latterly became a grime station. Transmitting from a broom handle across a 30 km city radius, these friends imagined were broadcasting to people like them (Nicolov 2017). Chatting on the mic, they pictured their schoolmates tuning in – the people they could see from the vantage point of the high-rise rooftop. Reflecting on his involvement as a DJ with the station, DJ Dugs explains, 'You could see your whole life from there, your community, your school, your friends' homes, your homes and the parks you went to' (Dugs and Woods 2016: 177) – the places and people intimate to you were projected through RINSE's transmissions.

This intimate relation with the community extended through jungle pirate radio's broader commitment to public service. With aerials spouting from public housing, it is not hard to make a material connection between jungle pirates and the welfare state. This is captured wonderfully in the 1996 documentary *Radio Renegades* which features DJ Brockie driving around East London pointing out a series of council blocks adorned with the tell-tale double stack aerials (Rooper 1996: 6.40 mins). But it was also the case that jungle pirate radio activated its own mutualisms. More so than any of the other sound cultures in this book, it was there to provide music for 'the massive' – that is to say, for public provision. And this was reciprocated, not beneficent. For listeners, jungle pirate radio was not just another broadcast on a crowded FM dial. It was real. It was a 'lifeline'.

> Pirate radio ... was so important to those of us who couldn't afford to go raving every week or were not old enough. It was a lifeline. [If Kool didn't switch on] it was like a relative had died. (Dugs and Woods 2016: 148–9)

The micro-massive

Jungle pirate radio extended the intimate presence of the dance floor to thousands of bedrooms and cars across the city, providing a common presence between otherwise isolated micro-environments. In a fractured and sped-up society, jungle pirate radio was a unifying force through which listeners in different places (physically and socially) felt they were known and heard by others. Its micro-sites comprised the 'junglist massive' whose co-presence unfolded through the airwaves.

> This one goes out to the Capital! (MC Shabba on Remarc and Flirt 1994)

In this simple and regularly heard invocation, the jungle pirate defines and calls on the collective, as they locate themselves in the massive – moving the city away from local identifications to a metropolitan culture set by the broadcast radius of the transmitters. These metropolitan presences were reinforced and maintained through phone-ins and radio call-outs to crews, areas and constituencies of the metropole. Listeners waited pen in hand to catch the station's number, 'held tight' for the impending shout to their driving/raving/bedroom/record shop/local collective, all the while hearing the other massives reeled off like a speed-regurgitated A to Z.

> Hertford, Bounds Green, Total Music, Dulwich Massive. Wicked reception in that part of town. Stratford massive, Ilford massive. Picking it up good over the east region. Leyton, Tottenham, Ladbroke Grove, Hackney, Clapton, Brixton. Big up to all the driving massive all around the capital. Kool FM one family. (Extracts from Marley Marl and Footloose on Kool FM 1995: 34 mins)

> Bigging up Antony and Amy. Bigging up Nicky not forgetting Danny. Bigging up Charlie. Bigging up Louise. Bigging up the Hendon massive. Bigging up Porsche, Denver, Isla, the Camden massive. (Extracts from DJ Zinc and MC Rage on Eruption FM 1995: 11 mins)

These shout-outs were untidy and error prone; imperfect, redolent of stutters, mumbles and background noise; the voice sometimes barely audible over the music. This patterning generated presence. Affirming the messy vernacular of everyday relations, it opened listeners to the station and to each other. This was not the disembodied voice of BBC unity (Loviglio 2005: xix). The MCs and DJs were people like you. This was 'a real interaction and it was worth something' (JME on SBTV and All 4 2017: 20.14 mins).

The intimate presence of jungle pirate radio was generated *with* not *for* the public. Regulated radio's dilemma of how to whisper into someone's ear while finding the tone for the every-person was not a dilemma for Dream FM and their like.[10] 'I always felt I was one of many thousands that would be listening', explains Dream

---

[10] See, for a discussion on regulated radio, Crisell 1994: 4.

FM fan Finch. 'And that's a feeling of inclusion and connection with all these people you don't know' (Finch quoted in Hancox 2011).

Jungle pirate radio presence was at once with the massive and with the city, but it was also personal and private. After all, jungle pirate radio still permeated the (intimate) spaces of private homes, bringing music into people's bedrooms and into their lives, intimately connecting MCs and DJs to their public.

> These guys Fize and Swifflee (on Dream FM) really brought hardcore and rave music into my bedroom and into my life. And I don't know what it is, maybe it was the style they did it in or the music they played, but it connected with me. (Michael Finch in Jackson 2011: 9 mins 58 secs)

This intimacy was distinct from that of regulated radio.[11] It was private and personal, yes, but it was not the passive pervasive wash of ever-on shoebox transistors situated across parental homes. Jungle pirate radio was an active experience. For the 'girls in their bedrooms' getting ready for a night out, listening because they could not go out or indeed switching back on in the early morning to facilitate their come down, jungle pirate radio had a dynamic relation to their lives. Indeed, it was even difficult to listen to it passively. Its demand for attention made drifts from background to foreground fewer and further between. Jungle pirate radio was never quite assimilated into the dominant aural regimes of home or city. Its vibration was rarely idle and was frequently intoxicating.

> Shout out to the girls in their bedrooms! (Rush et al. 1993: 43 mins)

Junglist knowledge

On a mid-1990s summer night, MC Top Cat reminds us, 'if you don't go to jungle college you can't get the junglist knowledge'. Junglist knowledge was an intimate knowledge; an insider status and posture, 'hard stepping into the 1990s' (DJ Zinc and MC Rage

---

[11] See, for a discussion of regulated radio as intimate to private life, Dyson 1994: 179; Loviglio 2005.

1995: 13 mins). It was forged through common experiences of crap work, the city, low quality council and rented accommodation, the downbeat fag end of Tory Britain, a disregard for police and authority ... and raves.

Jungle pirate radio collected and sonically distributed this knowledge in lock-step with the city. This knowledge was a plural wisdom. From Harlesden's reggae vibes and Isle of Dogs' hardcore, to the Americana and hip hop in Hertford and Stevenage, 'many music's [made] up jungle' and many knowledges, too (Marley Marl and Footloose 1995: 26 mins). Jungle pirate radio was an impressionist's map of the city; an affirmation of the heterogeneity of British underground wisdom, not property of the Island Nation.

While house music offered a kind of confetti reality, all dress-up, fun and excess, and acid techno afforded a counter-culture reggae-skank-meets-white-crusties-in-the-city-from-the-shires, jungle's knowledges were of city pleasure and glitchy darkness. All inner-city grit and suburban aggression with grins, jungle informed what it felt like to live in London and the south-east of England at that moment.

The name 'jungle' contained its own black diasporic wisdom – Duke Ellington (Jungle Sound), James Brown (Jungle Groove), Bob Marley (Concrete Jungle) and Grandmaster Flash ('It's like a jungle sometimes, It makes me wonder how I keep from goin' under'). It cited an area of Kingston and the feeling of black urban living in the US. When Congo Natty as X Project used the Jamaican yard tape sample 'alla junglists' in his 1991, it spoke an old but new knowledge to Britain, too.

Jungle's sonics and samples referenced the social issues of the time: drugs, inner-city violence, loneliness and disconnection (Ferrigno 2008: 94). Its MCing patched the Jamaican gunslinger over the Wild West to make sense, not of UK gun culture, but of a popular Atlantic rebel masculinity surrounded by the shock and awe of social violence. Its commentary channelled Martin Scorsese, the capitalist powerhouses of Michael Jordan Inc. and Premier League football in celebrations of snatch and grab capitalism, entrepreneurship and the merits of consumer society, for the boys.

These were combined in junglist knowledge. Jungle pirate radio understood the blues of a February winter and a mid-week lull at work. It knew the euphoria of the summer, the highs of Friday night and the morning after. In a wonderful piece of radio from early morning summer 1994, DJ Sparkie on Freedom FM reflects that wisdom of the dawning city.

> Sound of the Freedom on the FM 89.6. Shout going out to all the DJs out there. What's going on? Make your way to the studio! Coming from myself the Sparks. Shout to the DJs. DJ Vision if you're out there stop eating your cornflakes and get your bum down the studio. Shout to the Vinyl, all the crew. Get up the studio, you know the koo ... Sounds of myself the Sparks, been seeing you through for the last two or three hours. None of the DJs have turned up. All DJs hurry up and get 'round here. That goes out to the Extreme, the Darkzone, all the Mini crew, all the ex-Park Crew. Make your way to the studio. That's coming from myself the Sparks. Especially going out to the Vision. Get your cornflakes down and sort it ... Oye oye! That goes out to all the crew. Shout to the DJ Vision and his cornflakes. Let's go ... Shout going out to Wayne and all the other crew over Upminster, Church, that way. Shouts to all yous lot if you're locked in. Coming from myself the Sparks. I'll be seeing you later on. See you later. Shout to everyone. Hold tight! (DJ Sparkie 1994: 35 mins)

Sparkie's words and intonation are wisdom capturing the feeling of the city in those early hours. Sweaty skin and cold air, tired eyes and early light, ringing ears and bird song, fatigue and confusion. Cornflakes, alarm clocks and property flows suspended, as your alternative knowledge of your city is served back. DJ Spark's got you!

Multiculture

These presences and knowledges catalysed a form of multiculture whose vernacular rejection of Island Britain was pointed in its mundanity. Jungle was redefining everyday British culture on black terms (Koshick Banerjea in Pilkington 1994 (12.23 mins)), but more subversively still jungle was a multi-ethnic undertaking, confounding the very rules of racial order and governance (Congo Natty in Seely 1994: 11 mins). Jungle helped the multi-ethnic city hear itself.

For me when I grew up in the early 80s, on the street my friends were black and white and at the weekend my white friends went to the white clubs and by black friends went to the black clubs and neither would cross. There was obviously the odd white guy that would go and the odd black guy that would go. The friends that I grew up with was 90% black. My stepdad was black. My sisters are mixed race ... I used to go to those blues dances and promote for [Four] Aces ... I didn't see it ... but none of my white friends would come. When the rave scene came that's what appealed to me. I could go somewhere where my black mates would go and my white mates, and that's the beauty. (DJ Hype interviewed in Cordell 1997c: 5.06 mins)

Jungle pirate radio stations reflected this. Photographs taken at Kool FM in 1994 and 1995, show MC Remadee, Smurff, DJ Trace and DJ Ice, Blusey G, Skibadee and Tekka, Smurff, DJ Ash and DJ Slippery – black, white, mixed race, Turkish and Asian – in convivial repose (Clifton 2015). As Shy FX and UK Apachi's 'Original Nuttah' showed, it was now possible for a black British producer and British Asian MC (of mixed South African, British, Arab and Indian heritage) to rhyme in cockney, Jamaican patois and ragga, at hardcore speeds, over hip hop breaks, with US film samples and dub basslines, and sound like London with only minor concerns for cultural authenticity (24H Canal + 1994).

These forms of conviviality were endemic to the rave. At One Nation's New Year's Eve party 1997, MC Dett explains, 'One Nation, especially with being in London and everything. It's very multiracial, white, black, Chinese and Indian here. That's why it's called One Nation' (MC Dett interviewed Cordell 1997c: 0.18 mins). The open sartorial codes encouraged these movements. E's reduced social tensions. That didn't mean that racism evaporated, but in the rave there was a pervasive sense that it didn't matter as much who you were or where you are from, 'if the music's running, its running' (Zinc interviewed Cordell 1997b: 2 mins).

The sonic condition of jungle pirate radio, and the wider culture, was key to these intimacies. Whereas E gets some plaudits for the convivial co-presence of the scene, jungle's sonics held more sway still. This was partly on account of the music. The hybridity of jungle helped people from different walks of life hear themselves and each other. But importantly, too, the sonic condition of jungle means that they couldn't always see each other at the same time. Picked up aurally through radio, records and raves, racial markers (like skin colour) were not easily drawn from the sounds surface. Evidently, jungle was a black sound. No one disputed that. But it was a black sound that could not easily be closed down by white identifications of the black Other. In fact, the sonic condition of jungle ensured that any such act of racial authentication was error prone, and that disruption was celebrated by black and white producers alike.

> People can't say jungle's a black thing, they can't say it's a white thing. I could play you twenty fucking tunes and ask you which colour person made it, and you wouldn't be able to tell me. (Goldie interviewed in Taraska 1996)

From craft to DIY

> It was the excitement of it all, building something with a few transistors, a car battery, soldering iron, buying cables, making our own aerials, setting it all up, getting it to work and then listening at home. (Belle-Fortune 2004: 223)

Jungle pirate radio was more DIY than craft. The continual threat of regulator's raids mitigated against avenues of refinement. Regularly removed transmitters, both expensive and laborious to replace, ensured an emphasis on pragmatism. While the radio technology remained amendable to craft – indeed, the development of radio owes much to the tinkering of amateur enthusiasts[12] – for most pirate radio stations the emphasis was less on tinkering and more about staying on air.

This shift from craft to DIY is also culturally explained. By the mid-1990s the budding junglist had an array of sound and music consumables available to them. Affordable JBL and Cerwin Vega PA sound systems obfuscated the need to build your own boxes. Bedroom studio equipment provided the means to sample, sequence

---

[12]See for discussion, Douglas 2004: 15–16.

and engineer for a fraction of the cost of a professional set up.[13] There was a general shift in music culture away from craft and towards consumer culture. The intimate and expressive relations reggae sound system personnel had with sound technologies were contingently being displaced by the ready-made objects of late-modern society.

This is not to say that all features of craft evaporated. In pirate radio, the knowledge of transmission was still acquired and handed down through apprenticeship. The illegal condition of pirate radio ensured this by denying information and property streams access to that ambit of cultural life. Jungle pirate radio's frenetic game of cat and mouse with the authorities then ran alongside a slow and patient learning that supplemented pirate radio's wider patterns of mutuality and communality. This knowledge was also not property in the sense that, while different stations competed for sites and FM frequencies, knowledge of transmission was also shared. Mister T, for example, took technical and logistic experience from Passion FM and Rude FM to Origin FM (Belle-Fortune 2004: 222).

These tensions between mutualism and capitalism in some ways defined jungle pirate radio, and indeed wider society at the time. Jungle pirate radio was not a shop window. DJ's tracks were rarely identified, and DJs were unrecognizable off-air. Un-cleared samples – like the copied tapes of listeners – were widespread, denying liberal ownership in copyright form. Pirate radio itself was free. Its pleasures did not depend on spending power. Jungle pirate radio ran advertising[14] and charged DJs and MCs subs to appear on the station. However, this was not profiteering. The pirates on the whole did not make money from the stations and any revenue raised generally went back into day-to-day operations (Doran 2014).[15] And also, capitalist imperatives were never far away. By

---

[13]See, for discussion of bedroom studio consumables, Belle-Fortune 2004: 42; Ferrigno 2008: 78.

[14]The 1990 Broadcasting Act, which made advertising on public radio illegal, meant that a lot of this revenue came from the advertising of club nights, and services, such as tattoo parlours, that were illegal to advertise on commercial and state radio.

[15]Larger stations like Kool FM could sustain themselves and a few key people; and tied into record labels, clothing and clubs, they could generate bigger profits.

the second half of the 1990s jungle was becoming a viable business. When, in 1997, Guy Richie used EZ Rollers' 'Walk this Land' on the blockbuster film *Lock, Stock and Two Smoking Barrels* that call became louder. Now, creative industries' capitalism styled the offices of L. T. J. Bukem's Good Looking Records; and Bukem, Roni Size, Goldie and a few others were, and remain, recognizable celebrities (Modern Times 1996).

Hype

> Listening to pirate phone in sessions like this, I felt like there was a feedback loop of ever escalating exultation switching back and forth between the station and the 'ardcore 'massive' at home. (Reynolds 2008: 234)

Hype flows through the ether. Hype is presence, felt as a connective closeness, it is the affective modulation of jungle wisdom, and it is wholeness, a junglist disposition deeply and commonly held.

Hype moves in waves, heightening and diminishing through a multitude of interactions in the sonic, social and technological ecology. Hype is composed of many strands. It is comprised of the buzzes of different activities and emotional states. It is generated through technicians' search for transmission sites, climbing onto roofs and the risk and anticipation of broadcast; waiting all day, day turning to night, invested in the expectation of the future energies of the city. Hype is that 'something extra' or 'edge' that arises from those states (DJ Brockie interviewed in Rooper 1996: 3.40 mins; King Yoof quoted in JamieS23 2014).[16]

> With the authorities and DTI, it's like a cat and mouse game and it's a buzz when you creep about, sling the aerial up and eek around. (Eastman interviewed in Cordell 1997a: 1.20 mins)

> I have had it quite a few times that I lost everything, because they took all of my equipment. But that is ... well, you have to

---

[16] As DTI activities intensify, they then co-produce the buzzes that feed the hype. When they let regulation slip at the weekend, the release (combined with the catharsis of leisure time) amplifies the hype, in circadian pattern (Dugs and Woods 2016: 170).

experience it, I find it hard to explain. It is the excitement: are they coming tonight or not? (Frank, 43, cited in van der Hoeven 2012: 932)

Jungle pirate radio hype is crystallized not in the moment that the bass bins warm but after 2 hours of set up in a static buzz – white noise. The white noise collects the labour, the anticipation, the emotion, as it also sends it out to the city. The white noise is the first interface of ensuing public fervour.

There was nothing better than getting up there, plugging everything in and hearing that 'sshhhh' – that white noise. It's always an amazing feeling. (Eastman cited in Clifton 2015)

Hype is generated through the DJ's experience, the excitement of being called into play, feeling like you 'just won the FA cup', getting ready to play, going to the stations, hiding your records, covert, and then getting on the decks, and playing all that back to the city (DJ Target 2018: 34–5).

Back in the dance the hype is ferocious and funky in equal measure. It's a 'warrior stance' charged with 'militant euphoria' (Reynolds 2008: 239); it's 'exuberance without the gaiety, pleasure without the grins' (Belle-Fortune 2004: 94). That hype is fractured, explosive, deep and guttural, swinging, strutting, bouncing, momentary highs and dirty lows. There the Roland 808 digitized by Rebirth[17] and the manipulation of the Akai S1000 sine wave keeps the hype down below the bottom-end of Studio One – huge and booming, window rattling 'rumblizm' (pers. comm. Misinformation; DJ Nicky Blackmarket quoted in Reynolds 2008: 243); reggae's rebel remade as frenetic drummer, all speed and shapes, heating up the funk and burning out its warmth (Reynolds 2008: 241).[18]

Home after a night out, hype is tuning in, phone-ins, feeling the rave in the living room. The rush from the bass of a stereo that doesn't have bass. Your body recalling the dance through aural

---

[17]Made by the Swedish company Propellerheads.
[18]Winston Brother's 'Amen' break, 'Apache' from the Incredible Bongo Band, Lyn Collins 'Think', among others were prominent funk breaks used in jungle. See for fuller sound repertoire of these, shouldermove 2010.

signals that are no longer registered on the surface of the skin. The skin prickles, hairs on end with the memory of sensation, its intimation and its pleasure renewed ... then back into the ether.[19]

> Bwoy it jus' tek you. It's time to pump up the sound and metamorphosis with the world, gwan. We're all in cars, living rooms, bedrooms screaming for the rewind. (from Belle-Fortune 2004: 93)

> When we got home [from the club] we would switch on pirate radio and dance all night in Kemi's kitchen then go to work in the morning. (Storm on Rage cited in Collin 1997: 242)

As the ravers took the hype of the dance floor to their front rooms, so too did the DJs literally carry the hype from the rave to the radio studio.

> If DJs tested a tune they liked at a rave, and it got a good reaction – if it created a buzz – they would take it straight to the radio, still buzzing themselves to continue the vibe through that medium. (Nicky Blackmarket quoted in James 1997: 50)

On the radio the MC is now hype's principal conductor. Back on Kool FM with MC Det ...
Det sends a shout out to 'all the massive locked in ...'. DJ Brockie drops the intro to '16 Track Ting'. Det continues, '... especially those who voted us best radio station of 1995, watch the ride. The original style. Shout out to all those who are getting cotch time to stick a tape in ...'. The intro is building. '... Maximum moves out to the guvnor. Hold tight Smurff. Goes out to the Ragga Twins. Hold tight Flinty. Hold tight Deeman ... Absolutely rolling 94.5'. A sample signals the drop. Boom! Riding that sensation, moving lower, Det's toast flows between drums and bass. 'Steppa in a place time to brock wild, listen DJ Brockie with the true jungle style. Step back in a place, my DJ! ...'. Det moves out as the

---

[19] Jungle pirate radio is part of a network of bass delivery systems; it is not, as Goodman would have it, a material experience of the 'full-body-ear-drum' but is its recall (Goodman 2007: 52).

sample accentuates the bassline. 'For the ladies for the guys, big tings are gwan, as were nicing up the land, big tings are gwan ... For the Sunday night slam going out to all those that just don't give a damn'. Snares. The next break. 'Waking up your spine'. Det switches to double-time, riding the drums. 'How much more can you take?' (DJ Brockie and MC Det 1996, start).

Immediacy and impermanence

> Keeping it live in London city. (DJ Zinc and MC Rage 1995: 18 mins)

Immediacy and impermanence characterized jungle pirate radio's intimate depths. Jungle pirate radio was a live medium comprised of unique experiences not heard again. It had a feeling of ongoing precarity, of a system at the limit and on the limit. Listeners talked about being 'locked in' to the here and now, but in jungle pirate radio 'locked in' never implied permanence.[20]

Impermanence was the condition of jungle pirate radio (and wider society, too). Would the DJs show up? Would the authorities? Would the transmission fail? Impermanence was fed into the ether again and again by routine shout-outs seeking verification of the radio infrastructure.

> Please state your destination and location as we would like to know the reception quality in your area ... state your location or access will be denied. (Marley Marl and Footloose 1995: 8 mins)

MCs embodied the twin conditions of immediacy and impermanence, their staccato commands giving them vocal form. 'Short and sweet that's how we like it' (Marley Marl and Footloose 1995: 8 mins). This was matched by the DJs mixing practices, chopping and mixing records into a collage, always on the edge, always on the move, the anticipation and arrival of the next break, and the next and the next. If reggae and dub moved you out of time through the longue durée, jungle was of the here and now. It broke the mould of

---

[20] See, for discussion, Christoloudou 2009: 191.

daily existence but not through the expansive induction of trance. At nearly three times the speed of a resting human heart rate, it lit a fire under a single point in time.

For those 'locked in', this was all compounded by the precarity of the listening experience. This concerned the fast-moving nature of the scene. Some tunes were played once and never heard again. But it also concerned the transmission and reception technologies, the properties of the FM signal and the problem of brick walls. Listening experiences were always immediate – in the sense that you might never hear the tune again – and impermanent, because you listened in the knowledge that reception could suddenly be lost. 'It was always painfully frustrating, when you could hear a great set going on under the fuzz' (Mike Finch quoted in Hancox 2011). Most pirate radio listeners spent a good proportion of their time walking around rooms aerial in hand, trying to find that illusive sweet spot.

This unpredictability was central to the listener's relationship with the station. The impermanence of the jungle pirate radio listening experience was an intimate material connection to the radio itself. Odd as it might seem, the frustrating experience of moving round a room, antenna in hand, trying to clear the fuzz, was an affirmation of pirate radio's poesis and your place within that. Deeply frustrating and energizing in equal measure, it was elemental to the production of anticipation and hype that the dependability of regulated radio can never achieve, and nor, too, the OFF|ON qualities of its Internet versions.

The sonic intimacies of jungle pirate radio

From the early to mid-1990s jungle pirate radio was the metronome for London and much of the South East, too, a popular culture with wide reach, riven with contradictions. Co-presence and mutuality intertwined with capitalist individualism and consumerism; shared knowledges of welfarism and multi-ethnic conviviality were channelled at hyper-capitalist speeds.

Jungle pirate radio attained a condition of wholeness and depth known as 'hype'. Hype was illegality traced through buzzes, edginess of not-for-sale jungle culture. Hype was immediacy and impermanence laced dialogically through the rave, the radio, bedrooms and cars. Hype operated at speeds and rhythms beyond that moment's bourgeois capitalist myopia, providing for different imaginations of life, without actually proposing any.

Hype was not the authoritative garnering of the masses. Rather, its was predicated on co-production, through the call and response of the rave and the radio, and even the co-production of the technology as you wandered around the living room aerial in hand. Hype was horizontal, a sense of political belonging born of shared labour and a far cry from the increasingly hollowed-out democracies of that post-war period. Hype was not anodyne either, but immediate. Hype's immediacy demanded attention and investment that was wholehearted and deep, but also as precarious as the working-class city in which many ravers lived.

Jungle pirate radio's social intimacies of community and service extended from the pirate radio crews, to the neighbourhood and through the listeners, into the city and back. They provided systems of pleasure and support that mitigated the social atomization of the time. They were not the familial forms of reggae sound system, rather, they were part of fragmenting social affinities and institutions that continued to provide alternative networks of mutuality and identification.

These different scales of social and spatial intimacy were grounded in the materiality of the rave. As with the reggae sound system, jungle pirate radio was physically consecrated in dance. Its alternative cultural politics are otherwise unimaginable. Through pirate radio, the rave was extended across the city, connecting the otherwise disconnected late modern micro-environments of bedrooms and cars. Spilling out from the rave, the sonic transmissions of the radio then created a tangible multitude with shared practices, habits and ethos that attained a feeling of wholeness on a city scale, working around the otherwise visual racisms and atomized confines of daily life.

The multi-ethnic make-up of the city and the diasporic formation of the sound constituted a culture in which racial affinity and the racial ownership became estranged. Jungle remained a black diasporic music culture, but one that was composed by and for the multi-ethnic city, in which indifference to difference was routine; where ethnic and cultural difference could habitually lived and acknowledged but not always determining.

The sonic condition of jungle pirate radio allowed for contingent diasporic and heterogeneous wisdoms within a black dance music format to circulate. These were alternative to the dominant composition of racial modernity vis-à-vis black criminality and Island Englishness of the time. Through the sounds of jungle pirate radio, multiple marginal and popular knowledges travelled, at the same time as racial identification, self-curation and celebrity culture was frustrated.

This was patriarchal. Certainly, there were many female ravers and pirate radio fans, but at the same time its principal producers, practices and narratives were masculine (with the notable exceptions of DJ Rap, Kemistry and Storm). The culture as a whole – the risky heroics of the transmission sites, the cat and mouse territorialities of the city, the control of technology and airwaves, the steely warmth of the dance floor and, not least, the popular cultural references were all orientated to men, even if these were generally not sexually aggressive spaces.

Jungle pirate radio sustained intimacies in time, at odds with the normative proscription of the city. Its illegal condition, autonomous economic infrastructure and residues of craft allowed for expressions outside of corporate and state production rhythms. In jungle pirate radio, sonic expression existed as a way of life, not as a commodity, although property and personal brand was not far away. It was not a code for a different society. Jungle pirate radio was too synchronous with the capitalist mobilities of the moment for that to be the case. The ethos of craft and the longue durée of dub was after all replaced by consumer objects and communication speeds. But we don't need to demand that of jungle pirate radio either. It is enough that it was an alternative.

4

Grime and YouTube music videos

It's 2008. New Labour is burning out and the aspirational sloganeering they specialize in is bleeding into Tory austerity dogma. YouTube has recently been purchased by Google and the platform has expanded rapidly, becoming popular with young people and surpassing television as the most engaged with televisual medium for that generation (BBC 2019). Thousands of professional and amateur grime and rap videos are being uploaded, and YouTube channels like SBTV, LinkUp TV and Grime Daily are offering regular content. One of the first videos to make an impact in this format is Giggs' *'Talkin' the Hardest'* (Giggs 2009). Shot at chest height with a third-generation mobile phone camera, it is played back through countless computer screens and tinny speakers, quickly accruing tens of thousands of views.

'Talkin' the Hardest' is emblematic of the 'grimey' intimacy of that moment, but its YouTube mediation never sat well with grime's own mythology of council estates, record stores and pirate radio. The authenticity of the street seemed anathema to the everywhere-but-nowhere detachment of Web 2.0. Indeed, grime's archivists pay it only fleeting attention. But, as this chapter will show, grime did embrace YouTube and that embrace was as significant as the other sound cultures discussed in this book. As with reggae sound systems and jungle pirate radio, grime YouTube music videos tell a story of a moment in the transformation of black diasporic sound culture in Britain, one

in which autonomous infrastructures are absorbed by capitalist imperatives, treble supplants bass, hyperlocal presence is embedded, social media becomes the centre and the screen starts to sing.

Risky Roadz to SBTV

Black diasporic music videos were well established before the advent of YouTube. From the 1980s, home video technologies had found their way into sites of everyday leisure, including jungle raves and reggae dancehalls where fans documented musical memories for home viewing or commercial 'VHS tape packs'. Grime's first wave of producers and MCs were weaned on these offerings. DJ Target and friends watched Jamaican dancehall videos at Flowdan's parents' house in Limehouse, East London. 'My house was the arena for watching sound system clashes', Flowdan recalls, 'my parents had a load of videos and we used to watch them again and again' (Flowdan cited in Wiley 2017: 52).

Early grime music videos followed in these footsteps. By 2005, the Sanctuary in Milton Keynes was less well known for its Dreamscape and Jungle Mania raves than it was for garage/two-step/grime Sidewinder nights. The Sidewinder CD packs included DVDs featuring the likes of Roll Deep with MCs Wiley, Flowdan, Kano, JME, Riko, Skepta and others – names that became synonymous with grime. These packs were the stuff of legend and were consumed as much by those who went as by those too young to attend.

As CD and DVD packs flew out of record stores, MTV Base kept a steady stream of US rap and hip hop videos flowing into family homes. These hip hop videos – including music videos, behind-the-scenes videos, interviews and freestyles – inspired grime's DVD pioneers like Rooney 'Risky Roadz' Keefe.

> There were a few DVDs at the time, *Lord of the Decks* and the *Conflict* pack, but there wasn't something that showed the scene as a whole … So, I said to Sparkie, let's start a DVD. I asked my nan to lend me money for a camera, and the rest is history. (Rooney Keefe cited in DJ Target 2018: 145)

In this way, Rooney Keefe – along with A-Plus, Capo, Ratty and Jammer, and Crazy Titch – found himself at the centre of fast-growing grime DVD culture, selling products direct to record stores like Rhythm Division on Roman Road. These DVDs included documentary footage of pirate radio and raves like Eskimo Dance, but also included interviews, street raps and freestyles (Risky Roadz, and Practice Hours by A-Plus); clashes and battle videos (Lord of the Decks and Lord of the Mics by Capo, Ratty and Jammer); and behind-the-scenes footage (DVD with Aim High Volume 2 by A Plus).

In 2003, Darren Platt launched Channel U on cable TV, renamed Channel AKA in 2009. Channel U expanded the reach of grime, influencing the programming of public platforms like BBC 1Xtra and black commercial radio station Choice FM (Riki Bleu Music and Promotions Manager, Channel U, 2004–7 cited in Collins and Rose 2016: 153). Now, for as little as £100 production cost and a relatively small fee, artists were broadcasting their videos nationwide to an audience of millions, with the possibility of getting picked up by larger cable channels. Roll Deep's 'When I'm 'Ere' was played every twenty minutes on Channel U before finding its way onto MTV Base (DJ Target 2018: 185). Dizzee Rascal, Kano, Skepta, Tinchy Stryder, Tinie Tempah and Wiley all came to prominence through these means (White 2017: 72).

File sharing was well underway by this point. Napster, LimeWire, Kazaa and MSN Messenger had all been launched between 1999 and 2001, and these platforms permitted the distribution of garage and early two-step music files between desktop computers (Plastician cited in Collins and Rose 2016: 37). That culture was nurtured by early Internet forums like the RWD Grime Music Forum, which had 50,000 users. As DJ Logan Sama says, grime was the 'first genre in [the UK] to be born with the Internet, to be spread virally' (Logan Sama cited in Wiley 2017: 94). Between 2000 and 2008 these trends, and the draconian Wireless Telegraphy Act 2006,[1] started calling time on FM pirate radio stations, causing many to move online. Supplanted by CDs, MP3 downloads, LimeWire and iTunes, record shops and cutting houses were in their final throes. Many DJs had

---

[1] DJ Spooky discusses the implications of the Wireless Telegraphy Act 2006 (Gibbins 2015).

replaced their vinyl with more portable CDs and MP3s. 'I think the Internet's killed pirate radio', said DJ Logan Sama 'and I don't think it can come back from that' (Logan Sama in Mavros 2011).

As the videos for Wiley's 'Wot Do U Call It' shows, mobile phones were prominent in the early grime scene (Wiley 2004).[2] Indeed, the UK Garage crew Pay as You Go Cartel (that would morph into the grime crew Roll Deep) derived its name from an in-joke about mobile network errors during the summer of 2000 that made call-ins to Rinse FM free – 'It's the pay as you go show!' (Plague cited in DJ Target 2018: 71). Around 2006, mobile phones are fitted with better video cameras and become the technology through which grime videos are shot. The Sony Ericsson W810 was popular for this purpose. That model had a louder speaker that could be 'chipped' by young people to produce extra volume for backing tracks. It also had video and audio recording capacities. Come 2008, if you wanted to rap over an instrumental and record and upload it to YouTube, it was essential technology.[3]

Grime artists coming of age in this period did not then launch their careers on pirate radios or in clubs – the police's campaign against grime nights is well documented[4] – but through mobile phones and YouTube. Stormzy recalls:

> In secondary school, Year 7 [age 11]. I know a lot of grime artists started off on pirate radio, but I missed that era, I was way too young. I was MCing in the playground, spitting lyrics over mobile phones – Sony Ericsson, Walkmans, W810s, the Teardrop Nokia phones, all of that. Vital equipment! I never even had a DJ set where a DJ's playing vinyl and I'm spitting. (Hancox 2015a)

---

[2]This is also discussed by Zuberi 2014: 197–8.
[3]In his book, Richard Bramwell captures the audio version of this moment, through his discussion of the ways in which groups of young people share grime and hip hop MP3s on the backs of London buses, to create their own alternative public space (Bramwell 2015).
[4]During the tenures of Prime Ministers Tony Blair and David Cameron, grime music is criminalized on the basis of its blackness and becomes linked to an array of social problems (cultural degeneration, knife crime, violence). The introduction of the infamous risk assessment form 696 means that grime nights could be closed down on the basis of the music and the audience's ethnic designation as black. Grime responds to this with indignance. See JME for Noisey 2014; Bizzle 2006.

Young people still wrote lyrics in their bedrooms and maintained musical cyphers in public spaces, but by 2009 many more were producing and listening to music through their mobile phones and YouTube. Over only a few years, black diasporic culture had shifted its twentieth-century affinity for analogue sound technology to a twenty-first-century compulsion for digital and networked music videos. Social media was now the centre of black diasporic sound culture.

Grime's principal YouTube channels emerge at this moment. In 2006, sixteen-year-old Jamal Edwards starts SBTV (Smokeybarz Television). That is shortly followed by Link Up TV in 2008 and by GRM Daily in 2009. These channels follow the format established in the earlier DVDs: street raps (SBTV's F64 freestyle series is pivotal here – named after the 64-bar duration for the raps[5]), behind-the-scenes videos, documentaries and music videos of varying production values. By the time GRM Daily is launched, video content is not episodic, as it was for DVDs, but a continual stream. By the end of the first decade of the new millennium, SBTV is accruing millions of YouTube views and has become one of *the* sources through which grime is discovered worldwide (Jamal Edwards cited in Collins and Rose 2016: 170; White 2017: 77).

The (ongoing) crisis of neo-liberal capitalism

As New Labour took power in 1997, the free market ethos of the earlier Conservative administration was extended to a wholehearted embrace of deregulated global finance capitalism that would later result in the 2008 financial crash. Named 'casino capitalism', that business model became a central plank of New Labour policy, captured by Peter Mandelson MP's infamous comment that he was 'intensely relaxed about people getting filthy rich as long as they pay their taxes' (interviewed in Wighton 1998).

Despite early promises, these riches (filthy or otherwise) were not adequately redistributed to the working-class parts of the country decimated by Thatcher's restructuring of the industrial economy

---

[5]See introductory YouTube video for F64 (SBTV 2009).

in the 1980s. The discrepancies between Mandelson's rich and Thatcher's poor were nowhere sharper than in the London boroughs that birthed the grime sound. Haringey, Barking and Dagenham, Lewisham and Lambeth all have their place, but it was the former industrial East London boroughs of Tower Hamlets (particularly Bow) and Newham that were most important for the early scene (White 2017).

On either side of the River Lea, with the River Thames to the south, Tower Hamlets and Newham were multi-ethnic boroughs demographically composed through generations of migration. They counted with higher than average unemployment, lower than average pay, precarious employment conditions and multiple forms of social deprivation (James 2015; White 2017). They were also fringed with excessive wealth. To the south of Tower Hamlets was Canary Wharf's monument to global finance capitalism and Yuppie indulgence. To the south of Newham (on the other side of the Thames) was the Millennium Dome, New Labour's architectural realization of urban chic, aspiration and deregulation, that for a short time was to house an actual 'super casino'.

Grime's relationship to capitalism should be understood in this nexus. Grime is not anti-capitalist (neither was reggae or jungle); 'Money motivates me. I'm motivated by money', says Dizzee Rascal. Skepta hails grime pirate radio a successful 'pop-up investment' (Skepta cited in Hancox 2015b; Hancox 2013, Kindle location 249).[6] But nor were Dizzee and Skepta casino bankers or bourgeois entrepreneurs. After all, grime's 'hustle' was produced in the constitutive exteriors of those dominant capitalist cultures. Figured in the shadow of Canary Wharf, desiring its light, grime's early protagonists were always more certain that the money would run out than it would keep flowing. The presence of Canary Wharf in so many grime videos – for example, Shystie's 'Ima boss' and Upcoming Movement's 'Time is Up' (Ice Films Entertainment et al. 2011b; Shystie 2011) – was symbolic of these psychic and social relations, and indeed illustrative of how a scene so heavily invested in the myth of authentic street life did not only *not* reject wealth and celebrity as YouTube took over the autonomous infrastructures of raves and pirate radio, but embraced it.

---

[6]See for discussion of entrepreneurialism in grime, White 2017: 110.

Deterritorialized and hyperlocal

As YouTube took over grime, many of the social and spatial intimacies of the sound culture's pirate radio era were retained. After all, God's Gift and DJ Target were family friends, D Double E's and Jammer's dads had been at school together, and Tinchy Stryder, Dizzee Rascal and other members of Ruff Sqwad went to Link Centre youth club off Devons Road.[7] These relations went under various interpersonal stresses and strains, but they did not simply evaporate. At the same time, however, forms of presence were altered. As YouTube became more prominent and the media ecologies of pirate radio were displaced, people were no longer calling into pirate radios saying, 'Yes, Wiley, you're killing it!' or 'Big up Blah-Blah from Blah-Blah' (Wiley 2017: 82). From city wide call-ins and rewinds, the dialogues of the city were now increasingly online – simultaneously hyperlocal and deterritorialized.

Hyperlocality refers to the intensification, through social media platforms, of already existing neighbourhood localisms and is evident in many amateur YouTube grime videos. These videos perform hyperlocalism through the use of neighbourhood symbolism (local landmarks, gang colours and the male crew), which map onto territories and postcode areas. We can see this in the videos of the Newham group Upcoming Movement (Ice Films Entertainment et al. 2011a, b; Upcoming Movement 2012). These localisms are not new, but trace older rivalries between working-class neighbourhoods also negotiated through masculinity.

Through YouTube music videos these historical localisms are shrunk. Here, the junglist massive that opened out to the city, turns in. These hyperlocalisms are compounded by the young age and limited mobility of many grime protagonists, by the closing down of former sites of sociability – pirate radio stations, record stores, club nights and youth clubs – and by the policing of the remaining available public spaces – street corners, parks and stairwells. However, it is also the case that grime's condition of hyperlocality is shaped by the YouTube platform itself. Whereas a grime radio

---

[7] See DJ Target 2018; Collins and Rose 2016 for discussions relating to Tower Hamlets; and for discussion of Newham, White 2017.

pirate could stand on the roof of a tower block and imagine the community s/he was playing to, on YouTube these projections are unfathomable. Defined by plenitude rather than finitude, YouTube communities cannot be tangibly embraced or embrace you back.

These hyperlocalisms are compounded by deterritorialization. Deterritorialization refers to the ways in which social context, time and space are fractured and recomposed through YouTube. Whereas pirate radio relays through a central point (the radio station), the YouTube dialogue counts on numerous different dialogic partners. These multiple hyperlinked horizons, interlocutors and social contexts move as digital bits through networked space in several directions at once, speeding up and slowing down the arrival of the message; 'decomposing and recombining, multiplying and aggregating into different contexts' (Rubinstein and Sluis 2013: 30). As grime producer Tricks says, 'You can upload your stuff, send it out and anyone in the world can hear it', but importantly, and unlike Wiley's comments above on pirate radio, this *anyone* is an empty figure (Mavros 2011).

Relying on user inputs, 'aggregated practices of sequential viewing' are filtered and organized through algorithms written by global corporations to increase the consumption of online material (Airoldia et al. 2016: 1). These algorithms are produced far away from grime's local realities. Grime YouTube music videos then become nodes in a decontextualized network, where the sequence in which videos are suggested to viewers is determined not by the contents of a track or the call-in dialogues between DJs and listeners, but by the relative weight of the videos to each other (Airoldia et al. 2016: 4; Davidson et al. 2010; Zhou et al. 2010). In the absence of a city that can embrace you back, the hyperlocal then becomes an increasingly tangible horizon for deterritorialized expression.

For audiences, the hyperlocal appears as a key signature through which a deterritorialized grime video's credentials can be established. As East London producer Malacki da Monarch explains:

> When you see artists who are getting millions of views, you will find that most of their fans are not [from where they are]. If you go on Google Analytics, they are not from here ... they've got fans elsewhere. What people are catching on to is that it is very authentic ... people are intrigued, and that is commercial. (Interview with Malacki da Monarch, 15 October 2018)

Without requisite contextual knowledge, the concrete, tracksuits, crews, masculine somatic performance, angst and intensity function as open coded affective hooks, drawing the deterritorialized viewer in. Here distinctions between authentic 'hood videos and pastiche become blurred. SBTVs F64 road raps quickly realize the value of placeless localism, as do videos more obviously made for a global online market. Shot in the Barbican London, Skepta's 'Shut Down' takes the high culture brutalist aesthetics of the wealthy Barbican estate, to sell grimey urban locality to an online audience (Skepta 2015).[8]

Closed-circuit DIY

Through craft processes, alternative expression was built into the fabric of the sound system speaker boxes. Intimate knowledges askew to the dominant flows of capitalist production and information were grafted onto jungle pirate radio through illegal and pragmatic DIY practices. Grime YouTube music videos follow the DIY ethos of jungle pirate radio, in that they make pragmatic use of available technologies. The ubiquitous mobile phone provides young people with the means to play tracks and record music videos, and the YouTube interface allows for these productions to be shared online. However, despite some similarities, this labour relation is not of the same order as sound systems or pirate radio.

Mobile phones and YouTube technologies are designed with a low user skill threshold. Their nominally global reach, their profit imperative and their legality means they are manufactured, unlike sound systems and pirate radio, to be used by potentially anyone. There is no apprenticeship to be had and little alternative wisdom to be forged. Early grime YouTube producers like Jamal Edwards did not need apprenticeship from the likes of Rooney Keefe. Yes, you can 'chip' the Sony Ericsson W810 to boost power to its speaker, and this constitutes a minor hack, but it is negligible compared to the wider patterning of labour and expression through these technologies. Unlike the open knowledge and craft of the sound system or the wisdom of pirate radio, YouTube music video technologies simply

---

[8]This is not to state that these are the only forms of video possible in this networked media environment, but rather that there is a popular tendency in that direction.

require financial and physical access. With enough money and a few clicks of a button almost anyone can express themselves through its closed circuits.

MP3s, peer-to-peer software, social media platforms and streaming services such as YouTube still allow for the sharing of grime music, and therefore for a gift economy (Mauss 1966).[9] Discussed in some corners as a democratic musical revolution – freeing music from corporate ownership – what this actually entailed for grime was the mainstreaming of dominant capitalism through the alternate infrastructures, technologies and relations of the sound culture. As with reggae and jungle, grime had been sustained through its alternative media ecologies (not through major record labels) and through gift economies in which key instrumentals like 'Tings in Boots' by Ruff Sqwad (Ruff Sqwad 2003), 'Pulse X' by Musical Mob and 'The Matrix Instrumental' by Wiley (2010 [2001–6]) were shared and versioned many hundreds of times. As the alternative media ecologies of grime were displaced by YouTube, these forms of economy found affinity with the systems of prosumption that characterize labour exploitation and profit generation in Web 2.0.

Prosumption is a concept in media theory used to explain the ways in which the consumer of media products also becomes the producer of economic value, as their uploads, clicks (related to advertising) and personal information feed back through recursion systems into a platform generating value for a corporation, its shareholders and some monetized users (Fuchs 2014; Jordan 2015). As 'Tings in Boots' by Ruff Sqwad, for example, becomes shared through YouTube, the gift economies of grime then become imperceptible from the forms of dominant capitalism for which YouTube is designed.

This is a reorganization of grime's economy and contingently its temporality. This reorganization is happening across society, but the argument of this book is that for black diasporic sound cultures the effect on its alternative horizons is of particular concern. As grime videos are uploaded to YouTube, they attain the rhythm of digital capitalism. Here the analogue rituals of journey, purchase, selection and placing a record on the turntable is reduced to a

---

[9] A gift economy is one in which objects of value are given voluntarily, and in which the shared acts of voluntary giving form a reciprocal social contract.

click. Similarly, through the YouTube interface, the one-hour pirate radio mix or four-hour rave is returned to timeframes originally designated by advertising rationales. Kaplan argues that the length of music videos (four minutes) is determined by what MTV considered to be the optimum human attention span for an advert (the music video) to sell physical music in record shops (Kaplan 1987: 13). But four minutes is far more than the average movement between YouTube music videos. We rarely watch a whole YouTube music video, after all. Rather we flit between weightless offerings, bits of coloured light, decided by algorithms embedded in the side bars of our screens. We are less seeing/hearing the tune, hanging for four minutes until the next fix, than existing in continual throw-away frenzy of consumerist satiation deficit.

The same can be observed in YouTube's cannibalization of grime's sonic culture. As instrumentals and DVDs are ripped and uploaded to YouTube, their alternative time signatures are replaced. The alternative labour, creativity and expression of clubs, pirate radio and DVD culture are reformatted. The famous Conflict video is a case in point. This video documents the nascent grime pirate radio scene on the borders of Tower Hamlets and Newham. Shot at the Deja Vu pirate radio studio on Waterden Road, Stratford, it captures Dizzee Rascal clashing with Crazy Titch in the company of Wiley, Maxwell D, Lady Fury, God's Gift, Demon, Tinchy Stryder, D Double E and Sharkie Major.[10] Made by A Plus and released on DVD, the video becomes part of grime's insider folklore, only later forming part of wider public knowledge after it has been uploaded to YouTube. A pirate radio sound culture at odds with the dominant rhythms of the city (as we discussed in the previous chapter) and a DVD culture linked to those rhythms through autonomous record stores are then turned to the informatic temporalities of YouTube, as too in their media archive.

In this way grime artists growing up with YouTube and its surrounding media ecologies come to see the economies, social practices and temporalities of their scene through their intimacy with YouTube, and indeed see a YouTube music video 'doing the numbers', not the first pirate radio slot or sound system appearance, as the benchmark of success. This is not to suggest

---

[10] See for discussion of the video, Hancox 2018: 1–2.

that the majority of grime music videos are conceived in terms of selling music. While professional performers and producers work in these rationales, the vast majority are amateur productions and any monetization of the video is non-existent or negligible, but they nonetheless know themselves on these terms. The intimate relation between producer, technology, expression and knowledge is practised and narrated more through the regulated rationales of digital capitalism and less through the alternative economies, practices and temporalities of gift and craft.[11] This is a shift in the intimate relation between black diasporic music cultures and their technologies. The patterns of expression possible through sound technologies are increasingly manufacturer designated. You can record and upload the video that you want, but an intimate creative relation with its closed-circuit technologies is impossible.

Grime, coldness, agonism, intensity

Sonically, grime is a diverse music, about 140bpm, faster than garage but slower than jungle, and typically scored in 4/4 time with 8- or 16-bar cycles. It is largely comprised of electronic sounds, synth keyboards and some sampled soul and R&B melodies (as with UK Garage) and employs programmed drums and bass lines (as with jungle). Initially, this follows a two-step garage pattern filled in with jungle snares and hi-hats (derivative of hip hop) (Charles 2018: 5–6; Paintin 2016: 11). The bass lines draw from reggae/dub and the lyricism extends from reggae toasting and jungle MCing with hip hop and rap influences.

Those sonic properties correspond to the biographies of its producers. Skepta grew up in Tottenham listening to reggae, dancehall, Snoop Dogg, jungle and UK garage. No Lay wrote

---

[11]In the context of shorter and shorter attention spans, and for those who imagine success on financial terms, a continual flow of audio-visual output is required. When audio-visual stimulus takes precedent in this way, the content of the music becomes empty, reduced to an affect-garnering shell. Placed alongside personal brand and merchandising it becomes just another way in which visually organized business is done. On consumer terms, Boy Better Know as celebrity, brand and music video pastiche is a minor example of this.

poetry and jungle bars in her bedroom (No Lay 2011). As Footsie says, 'It's so deep sometimes I think I'm not doing nothing special, other than carrying on what was already done' (Footsie cited in Hancox 2018: 34).

Grime is a version of what came before but it's also its own story. Grime's recent history can be narrated through its development from jungle and its split from UK garage. Whereas UK garage was concerned with champagne, labels and loafers, grime's aesthetic was of street life, tracksuits and trainers. 'Grime is not about showing off', said Crazy Titch (Crazy Titch cited in Collins and Rose 2016: 35). It's not about 'flossing' (flaunting apparent wealth). 'What I do is really raw', said No Lay. 'It's straight to the point, no fucking about' (interviewed in Garland 2017). And while the 'grime' epithet caused consternation among its early protagonists who saw it as a journalistic homogenization of a plural and grassroots sound – a reduction mocked to great effect in Wiley's 'Wot Do U Call It?' (Wiley 2004) – it was also quickly adopted as an insider descriptor for 'What you see when you wake up in the morning … a lot of grime in the area, a lot of grimy things happening' (DJ Trend from East is East documentary cited in Hancox 2018: 18).

This is reflected in grime's early video culture. Released on Channel U in 2005, Kano's 'Ps and Qs' uses the roads of London as a backdrop. A cameo from The Street's Mike Skinner is all that remains of the high budget Missy Elliott-esq aesthetics of UK garage (à la So Solid Crew) (Kano, 2005). Roll Deep's behind-the-scenes documentary, *One on One with Roll Deep*, released with their debut album *In at the Deep End*, takes this aesthetics further, developing what becomes the predominant grime music video style (Roll Deep, 2005). As also evident in No Lay's video for 'Unorthodox Daughter', council estate locations, graffitied walls, chest-high camera shots and basic lighting become default (No Lay 2008).[12]

On the dance floor, the accompanying two-step sonics are described as 'dirty', 'mucky', 'grotty' and 'grimey' (Hancox 2018: 17). 'Wheel up that grotty tune!' calls Kano (Kano 2005a). This is an appeal to the soil of former industrial places. 'I come

---

[12] That video is shot on the Lansbury Estate, Bow, London amid underground garages and National Front (NF) graffiti.

from nothing – I come from the underground, pirate radio stations, *I come from the ground man*', says Dizzee Rascal in a 2003 interview for BBC News (Dizzee Rascal cited in White 2017: 15; my emphasis). Shystie's first album, *Diamond in the Dirt* (Shystie 2004), is not then just a play on 'diamond in the rough' but sings the sonic history of London soil itself. This is not just a rejection of ostentatious aesthetics but a claim to the material history of East London working-class life, its factories, soot and grease. It is a revisionist claim against UK garage, as it is also a reckoning with the clean glass of Canary Wharf, and the social violence of those aesthetics.[13]

As grime is a sonic document of the post-industrial working-class city it is also a sonic document of black Atlantic life. The ground and earth play, too, through reggae, jungle and, indeed, jazz. Duke Ellington's firm grounding bass, and the downward bass-heavy skanks of UK reggae and jungle dance floors become more intense, harsher and chromatic in grime. This is where D Double E meets Burial by Leviticus and Boy Better Know meets UK hip hop's Klashnekoff.

Grime's sonics are also cold. Towards the end of the 1990s, the RAM records jungle productions of Andy C, Optical and Ed Rush start to develop a molten metallic sound. In the hands of Groove Chronicles' UK garage productions, they become further chromatized – the tracks 'Black Puppet', '1999' and 'Stonecold' are exemplary here. Grime extends this repertoire through its 'sparse, glacial, machine-made beats leaking out of a thousand phone ringtones' (in-sleeve from Various artists 2005). Wiley adopts the pseudonym 'Eskiboy' and in 2002 produces two foundational instrumentals 'Eskimo' and 'Ice Pole'.

> The sound came from our situation. It's a cold, dark sound because we came from a cold, dark place. These are inner-city London streets. It's gritty ... Eski, igloo, ice, cold that all comes from my childhood. The pain, the isolation, the frustration ... being in a dark place ... My sister Janaya understands; from when we were kids, she's always been in the igloo with me. (Wiley 2017: 79, 255)

---

[13]I have discussed this theme in more detail in James 2018.

Wiley's igloo traces the *Invisible Man*'s hole.[14] The igloo itself, a metaphor for social negation, retraction, isolation and sanctuary, is the structure for the sonics of his music. As Ellison's warmth is to jazz, to darkness, to light, to endings and new beginnings in the context of a racist and capitalist society; Wiley's igloo is to grime, to whiteness, to darkness, to social wilderness and coldness.

Wiley's igloo is white but dark, and 'darkness', anger and pain (agonism) colour the sound of grime. Grime is 'about aggression and anger and pain', says Crazy Titch (Crazy Titch cited in Collins and Rose 2016: 35). Sonically, this 'can be hard on the ear, the beats can be disturbing and brutal' (White 2017: 30). As the in-sleeve to *Run the Road* puts it:

> The aggressive, minimal sound structures shaking out of cars at the lights all over the London and beyond; the tense soundtrack of stations like Rinse and Freeze that has re-energised London's pirate radio scene. The unclassifiable beats rolling out the door of record stores like Rhythm Division in Bow and the aggressive, spitting rhyme battles jumping out of DVDs like Lord of The Mics or Conflict. (in-sleeve from Various artists 2005)

Here we find the lyrical assault that is Lethal Bizzle's 'Pow!' 'made of pure frustration' (cited in Collins and Rose 2016: 23), Dizzee Rascal's *Boy in da Corner*, 'an astonishing debut, a projectile vomiting of anger and violence, alienation and confusion' (Hattenstone 2010). As sociologist Yusef Bakkali puts it, this is 'the pain of the mundane' (Bakkali 2019). 'The best MCs speak from pain', says Wiley 'from when they were down and never had a fiver' (Wiley 2017: 297). This is the anger of 'dysfunctional families from early when you are a grown man in a house from a youth', says Big Narstie, 'tearing down road to go Iceland [to put food on the table] and make sure your mum's okay' (BBC Radio 1Xtra 2015: 7 mins 20 secs). Grime sounds out neo-liberalism's 'fuck this. Fuck everyone' moment (Wiley 2017: 255). 'Everyone's

---

[14]Ellison's *Invisible Man* contains many references to coldness as inhumanity, and this is counterposed to the warmth of the dark, jazz-resonating hole explored at the end/beginning of the book (2001).

so angry at the world and each other. And they don't know why' (Wiley cited in Hancox 2018: 68).

In grime, these sonic expressions of grime, coldness and anger are intense and energetic. 'I was always so gassed', says Ghetto (NFTR 2015). 'All this violence around you can fuck you up, but as a musician it can help you. It gives you raw energy' (Danny Weed cited in Collins and Rose 2016: 27). Battle clashes, call outs, sending for people, MC culture and war dubs then compound an already existing social intensity, born of neo-liberal capitalism, marginalization and alienation in late New Labour and early Tory Britain.

Sound boy I can have your guts for garters, turn this place into a lyrical slaughter. (No Lay 2008)

There are numerous examples of this. Terror Danjah Ft (Riko, Bruza, D Double E and Hyper) 'Cock Back', a track about gun play, redolent of aggression and coldness coupled with speed and energy (Various artists 2005). Dizzee Rascal's staccato, fast pace and high-pitch attack on *Boy in da Corner* and in his sidewinder collaborations with DJ Slimzee. No Lay's relentless wordplay for 'Fire in the Booth'. In the rave this intensity shifts UK garage's 'one finger skank, two-step in your new shoes' to grime's 'put up your gun fingers and see who you might move to' (Kano 2005a). These registers echo Beefy in Babylon, Ms Dynamite and Lauren Hill, but so, too, the hard scowling sonics of jungle.

Which is also to say that, while the agonism of grime stems from internal pains, is also shared. Devlin recalls, how as a young man, the agonism of the new grime sound called him from his home in Dagenham outer East London; it moved across London, establishing connections out of isolations (DJDUBLTV 2017). 'I feel what I feel', said Ghetto, 'and it just so happens a lot of other people feel it too' (NFTR 2015). Indeed, one of Wiley's intentions was for other people to find themselves in his sound (Wiley 2017: 255).

Lo-fi

These sonic intimacies – grime, coldness, agonism and intensity – are retained and altered through YouTube. To approach the significance of this we can engage with what Monique Charles calls grime's lo-fi sound (Charles 2018: 7). Whereas 'hi-fi' denotes a listening experience that replicates, often through expensive technology, the mythical clarity of real and physical music, lo-fi is a listening experience characterized by distortion.

Grime's lo-fi sound is partly accounted for by the DIY technology used to make many of the early instrumentals – cracked versions of FruityLoops, Mario Paint for the SNES and Music 2000 for the PlayStation (used by Skepta, So Solid Crew and Dizzee Rascal) (Hancox 2015b; 2018: 61). Far from constituting a problem, in grime, the lo-fi sonics produced through these technologies are culturally appropriate to the sound culture's wider structure of feeling. As grime journalist Prancehall/John McDonnel notes, 'a lot of the early grime was actually quite bad production. I liked that' (Prancehall/John McDonnel cited in Collins and Rose 2016: 37).

This lo-fi sound is exacerbated by the YouTube media ecology. As noted, early YouTube street raps often used backing instrumentals played from mobile phones, with the MCs lyrical performance video-recorded back into another mobile phone for upload to YouTube (Upcoming Movement 2010). The final audio playback striped the original lo-fi instrumental of any remaining bass, further reducing definition and increasing the levels of distortion. In some cases, the original instrumental was only vaguely recognized in the final audio through its tinny rhythm and distorted mid and treble.

Video content goes through a similar process. Third-generation mobile phones' low resolution, poor quality lenses and insensitive handling of light create dark and grainy pictures, while the YouTube platform limits resolution (and length) through its 100MB upload limit (Gannes 2009).[15]

Treble and lyricism

The lo-fi sonic condition of grime YouTube music videos strips the bass from music videos, emphasizing the treble. In the previous two chapters we have attended to the special importance of bass to black diasporic music. We find this at its most instructive in the reggae sound system, in which the bass itself is part of the modulations

---

[15] After September 2008 this limit was increased to 1GB.

of affect and political demands. As we trace this into jungle pirate radio, we note that (powerful car stereos aside) we are in a bass referent media ecology – where the bass heavy rave is central to the radio playback experience. Grime is still bass music, part of bass culture, but as we enter the era of YouTube music videos, and as the raves are closed down, grime is less centrally defined by a felt relationship to bass.

Many early grime instrumentals are not even particularly bass heavy. In a DJ Slimzee set from an early 2000s Sidewinder event, Dizzee Rascal calls out 'listen to the mids', not listen to the bass (Dizzee Rascal and Slimzee c. 2001; 15.43). Indeed, as Hancox notes, 'The glorification of treble culture in grime reached a peak of forthrightness with the Slix Riddim "No Bass", rinsed by the likes of Ruff Sqwad, Bossman, and scores of mobile phone DJs throughout 2005/6' (Hancox 2009). However, this is more than a predilection for different frequency ranges because, as mobile phones and YouTube come of age, bass is also being actively tuned out of the productions.

Musical recording equipment has been historically tailored to the frequency range of sound reproduction technologies. The particular frequency ranges of LPs determined how studio sound was recorded and how microphones were designed (Marshall 2014: 54). The same is true of MP3s and mobile phones. Music coded into the MP3 format for playback through mobile phone and computer speakers likewise reflects the treble range of mobile phones.

This emphasis on treble tells a different story to that carried through reggae and jungle bass lines. From the moment at which cars were fitted with radios (1940s and 1950s), there has been an obvious affinity between treble culture, mobility and modern white sonics. This is a question of weight and weightlessness – the sonic and physical heaviness of black diasporic sounds giving way to the portability of white consumer movements. For those who place importance on the way bass conveys a 'moving/hurting black story' (Johnson 1980), the shift from heavy weight sound to weightlessness and treble culture, from the city resonant with sound systems and low end to one coloured by the hiss of mobile phones, is profound.

However, this is also only a partial analysis because the wisdom of black diasporic sound cultures is not only carried by bass. To suggest it is implies a reductive affinity between black diasporic life and only one range of sound frequencies. Fred Moten cautions against this by tracing treble histories in black diasporic sonics – in the snares and screams of Aunt Hester, Al Green, Michael Jackson, Max Roach and Abbey Lincoln (Moten 2003). In YouTube grime music, treble comes to the fore through the snare. As Richard Wiley (reggae musician and father of the 'Godfather of Grime') notes, the snare is the most important instrument in grime, not the bass. But it also comes to the fore through the human voice of the MC.

The rise of MC culture in grime is partly explained by the economic realities of young people involved in the scene. DJing is a relatively expensive pastime. You need decks, a mixer and a regularly updated supply of music. To MC you only need your voice and a microphone. So, as De La Soul rapped 'everybody wants to be a DJ', Wiley rapped 'everyone wants to be an MC' (for Wiley comment, see Hancox 2018: 52). As MC Det bemoaned the dominance of DJs in jungle culture, DJ Logan Sama occupied the backstage of grime events. This shift to the human voice does not, however, return grime's sonics to an Elvis track playing in a 1950s car on a US highway, in which the bass has been replaced by the banjo, because the whitened and lightened sound of Elvis is not the same (historically or culturally) as that of the grime MC.

In grime MCing, crews like Dreem Team and Pay as You Go Cartel broke from the whiteness of UK garage's front of house announcers, re-engaging a lineage of toasting that extends from reggae and through jungle. After all, nearly every MC in Pay as You Go had learned their trade in jungle (DJ Target 2018: 75). Maxwell D studied Stevie Hyper D's lyrical structure of eight-bar hooks and reloads. D Double E was a scholar of MC General Levy (D Double E and Maxwell D cited in Collins and Rose 2016: 30–7). That is to say, that as grime sonics are increasingly turned to the treble range, what is heard are both the mobility of digital capitalism *and* the reinvigoration of black British vocalism. It is a return of human soul in the age of the digital machine.

The sonic screen

The YouTube screen recalibrates grime's sonics, re-colouring the sound, re-sensitizing its affect and absorbing its sonic matrix into a visual field (see Berland 1993: 20). Through YouTube grime videos we come to see the sound through the video more than we hear the sound through the speaker. Music we used to know through sound (that videos made strange) comes to be known through videos (that makes audio alone seem strange). This, as Berland notes, is a form of 'cultural cannibalization', in which the [sound] becomes 'digested lifetimes ago ... consumed by the image, which [is] singing' (Berland 1993: 31).

The visual does not displace the sonic here because sound and vision are not neatly partitioned. The visual cannot be reduced to textuality (representation, symbolic, etc.) and does not simply overwrite sound, which is also to say that the whiteness of textual modernity does not erase the blackness of grime phonics. Visual culture is more than textuality, and therefore more, too, than textuality's historical affinity with the development of capitalism, racism and Empire. The visual is textual but so, too, is it affective and sonic. We can feel the warmth of an image and hear its shouts, screams and whispers. With music videos, then, image consumes the sound, but on the singing screen the textual has not displaced the sonic. Rather, the sonic is modulating to the affective energies of the visual field.

Unlike sound, which fills space and surrounds you, extending your body and soul into the social and historical, the screen surrounds itself with you, drawing your eyes to a single spot and fixing the rest of you before it. The ocular embrace is intimate and absorbs you into a pinhole of space. As Guattari notes on television:

> When I watch television I exist at the intersection: 1. of a perceptual fascination provoked by the screen's luminous animation which borders on the hypnotic, of a captive relation with the narrative content of the program, associated with a lateral awareness of surrounding events (water boiling on the stove, a child's cry, the telephone ...), 3. of a world of fantasms occupying my day dreams ... It's a question of the refrain that fixes me in front of the screen, henceforth constituted as a projective existential node. (Guattari 1995: 17; cited in McCormack 2013: 147)

The YouTube interface also re-colours the properties of sound, but not as the TV. YouTube is designed on active not passive engagement with the screen, and that further disrupts the ways we listen. In fact,

YouTube is not designed for listening at all. It is geared to watching, and to consuming with your eyes, at speed. The intensity of the blue light stimulates adrenalin in our bodies, raising resting heartbeats, inducing a chemical high that is addictive and compulsive. The habitual scanning of the algorithmically organized side bar for visual consumer cues, loosely absorbing the bright atmosphere of the track playing, before the compulsion to click takes over and the video is changed.

Through YouTube music videos 'we are captured doing not what we want but what we must', says media scholar Jodi Dean (Dean 2010: 21).[16] And this state of agitation compounds wider social anxieties redolent in neo-liberal society. Talking about the relationship between music videos and everyday life Jordan, a musician from Plaistow, East London explains: 'People are looking at little bits of information, taking this and that, and nobody knows how to stick to anything ... people get agitated ...' (Jordan King interviewed by Malcolm James, 19 November 2018). The agonisms of grime sonics that called across the city are then were reworked through an interface that specializes in rupturing deeper human connections, compounding our general state of social negation, calling it into action, rather than moving us beyond it.[17]

YouTube further heightens particular sonic intimacies of grime. The MC's voice is heard but now also seen. As the Conflict DVD comes to be known through YouTube, listeners no longer have to guess which MC is which, as they did with the original pirate radio broadcast. The recall of intonation, pitch, metre and content is a lost art. Those arguments held between friends over who is who on an audio track are now given visual clarity. The identity of the rappers, the racial and gendered codes of their bodies and their somatic movement start to sing, calibrating the hype and the intensity of their wordplay ... as they spill out onto the roof.

In this way, too, the open sonics of agonism, those that called to Devlin in Dagenham, are folded back into the visual and racial

---

[16]'Internet – inter-*net*, worldwide *web*. Why is everybody logging into a world wide web. Why do you want to log into a web. The only thing a web does is catch them, traps them. Everyone is trapped in the inter-*net*' (Jordan interview).
[17]See for complementary commentary on the affinity of networked media with racism's scavenger ideology, Titley 2019.

regimes they formerly exceeded. As the excessive belligerence of the police towards Giggs showed, the visual coding of *'Talkin' the Hardest'* stuck to Giggs's body in ways not possible when (as with jungle) blackness and (by racist association) criminality is an assessment of sonics alone.[18]

YouTube grime artists start to understand their expression on these terms too, learning social roles, performing social scripts for the camera. 'They learned to be in front of camera', says Keefe (Risky Roadz 2017). Classed and racialized performances of anger then become part of the repertoire of effective expression. Grime artists' performances become conditioned by the racialized affective economies of YouTube music videos. In this way, the possibility for the 'grottiest' tune, where sound comes before ethic affinity, is tightened but not eliminated – grime remains a multi-ethnic genre in that sense.[19]

The sonic intimacies of grime YouTube music videos

From 2008 onwards YouTube became the centre of grime's media ecology, displacing raves, pirate radio, record shops and cable TV. The screen-based dimension of this wasn't unprecedented, black music video culture was already well established in Britain, but the affinities between music video, mobile phones, Internet platforms and fans were new.

In terms of presence, while earlier relations of kinship and community remained, the intimacies of the sound system and the pirate radio were increasingly absent. Being on the dance floor or in radio call-ins with people like you was now secondary to deterritorialized dialogue. Verifications of yourself and others came

---

[18] The Metropolitan Police embarked on a five-year campaign to close down Giggs (Wolfson 2013).

[19] These videos, as part of black diasporic and working-class sound culture, are undeniably central to the ways in which Britain relates to itself, which is also to say that they continue to disrupt normative nationalist, capitalist and racist representations on mass media. While they may adopt some of the normative racial scripts, they are also nonetheless black, multi-ethnic and working-class young people representing themselves to the masses.

not through shared sonic knowledges of the past and everyday life, but through impressions left in view numbers, comments boxes and Google Analytics' data. These deterritorializations accompanied entrenched localism. In the context of pervasive neo-liberal negation, and in the absence of former sites of social mixing (record shops, youth clubs and city-wide pirate radio broadcasts), pre-existing definitions of neighbourhood insiders and outsiders were revived. This was a shift away from the city that embraces you back, and one that looked to local rivalry rather than metropolitan movement for comforts.

Although redolent with images of street life and local affiliation, as the autonomous hubs of pirate radio stations and the sound systems were replaced by algorithms, grime YouTube music videos were increasingly for elsewhere, for YouTube. Even those productions rooted in the street were consumed as pastiche outside of their immediate circles.

The alternative rhythms and expression of craft and illegal DIY were denied by the closed circuitry of mobile phones and online platforms. These locked technologies were proprietary, too. The form of mutuality and gifting that has long existed in black diasporic sound culture was co-opted into their systems of prosumption that capitalized precisely on the forms of cooperation that had formerly persisted beneath profiteering.

This was not at all unique to grime culture. Rather, it was shared across society, part of the broader shifts in technology, capitalism and neo-liberal cultural politics affecting all social strata. Grime YouTube music videos were then constituted with wider social forces, and not marginal instances of cultural life. It is also true that what was at stake in this shift for black diasporic sound culture was particular. The historical formation of black diasporic sound culture vis-à-vis regimes of whiteness, textuality and capital had occasioned forms of wisdom and social practice at odds with dominant culture, carried through sound. These sonic intimacies were porous and open in ways that white capitalist textuality could not be on account of its affinity with privacy, property and racist order. And, as such, they came to tell the stories of black diasporic life and, increasingly, the multi-ethnic city. The move to screens and digital property, while part of the transformation in wider society, is not evaluated in the same way because in black diasporic sonics there was something else at play, for everyone.

The previous two chapters discussed how sonic intimacy comes to be named. The reggae sound system's condition of wholeness is called 'vibe', for jungle pirate radio it is 'hype' and for grime it is 'grime' – lo-fi, grit, soil and the ground. Extending social and sonic life, grime video culture, too, was grimey. It expressed a deep knowledge of industrial and post-industrial places under erasure from the cleansing aesthetics and architectures of the Millennium Dome, Canary Wharf and latterly the Olympic development, and it conveyed a black diasporic wisdom concerned with the ground and with bass.

Through grime YouTube music videos, the aesthetics of griminess remained but the felt story of bass was engineered out. Digitally weightless grime music videos were then aesthetically grimey, but with close affinity to capitalist white mobility. With grimey aesthetics retained and the bass tuned out, treble – a sonic ally of mobility – returned in the form of the MC's voice. These vocals were grimey, too, faithful to the 'fuck you' neo-liberal capitalism of the moment. They were expressive of the shared depths of agonism and energy between like-minded people. Conveying echoes of rage, they provided living resources for alternative mutualisms without any other particular demands.

Seen on the screen, sound's openness to the city was warped. YouTube music videos returned multi-ethnic and convivial sound cultures to registers of race and racism. Who best conveyed rage to sonic affect was again visibly a question of skin colour, of the racial coding of rage and energy through the screen. Likewise, the blue light of the monitor, the intimate close-to-face interface and the compulsion to click re-coloured those sonic mutualisms, altering its properties and the stories it conveyed. The coldness of the sonics and the coldness of the screen, and the agonism of society and that of social media mixed, turning resources for alternative life towards never-satiated activations of pressure.

5

Conclusion: From left critique to alternative cultural politics

When Nadia Rose performed live at Village Underground in London in 2017 to a soporific room, Snapchat gave her love. Turning from the audience to the rapping circle and social media, she shifted through different modes of presence. From the cold crowd, she recharged in the warmth of rounds. From the warmth of the rounds she reached for the tactility of digital media. The room, full but empty; the cypher, deep and proximate; the social media, infatuating and never-satisfied – of the three, the cypher and social media were in dialogue. The former supplementing the deficiencies of the latter, the latter requiring the former. But as the live audience receded into the background, social media was in ascendance. Around that turned the unfulfilled love of one evening in Shoreditch.

To say social media 'gives us love' is to employ a platitude. That platitude makes sense only because we really do feel that social media provides something akin to love – at least as we understand love in the contemporary transactional sense of reciprocal self-affirmation. To say social media gives us love is to then confirm that the deep intimate potential of human relations has been reduced to screen-mediated consumerism. It is to signal the ways in which the affective environment of social media replicates those feelings. To say social media gives us love is to say it not only replicates love but provides it freely and easily. In a society in which even transactional love often feels hard to come by, to say social media gives us love signals infatuation with what social media provides readily. As with all transactional love, social media love is never enough.[1]

To say social media gives us love provides a window onto changes in sonic intimacy. In the context of this book, to say social media gives us love is to acknowledge the ways YouTube grime music videos provide for an intimate transaction; a transaction that is never complete, that reinforces social negation; a transaction that finds less traction in the collective of the dance floor than it does in multiple verifications of self; a transaction that moves more to the affective aura of the networked screen and restless interface than it does to the haptic rhythms of the night; a transaction so overwhelming one needs to hunker in the cypher so as to not be pulled apart. But listen! Hunkered in the cypher we find an embrace that is not dead, that is still vital. The cypher's embrace is mutual, democratic and horizontal. As it bolsters the deficiencies of social media, it affirms the necessity of sonic mutuality for life.

One night in Shoreditch draws our analysis to a close by opening a discussion on the kind of scholarship we might pursue; where endings are beginnings pointing to forms of evaluation we might better avoid. This final chapter begins by making a case for the approach to cultural analysis adopted in this book – for why a relational approach to cultural analysis is important. That approach has provided insight into the alternative cultural politics of black diasporic sound cultures. This chapter ends by evaluating the sonic intimacies of those and by clarifying what is at stake in them.

It matters, after all, what questions we ask, how we ask them and how we reflect on the answers. As authoritarianism, nationalism and racism coalesce powerfully in popular culture, in the UK and globally, our attention must be resolute. Do we provide latitude for alternative flows of life, or do we unintentionally re-inscribe the same assumptions from which authoritarianism, nationalism and racism spring? That interrogative may seem abrupt, given the appraisal that follows is of critical left scholarships, but it is nonetheless necessary. Because while critical left scholarship is

---

[1]See, for wider dialogue on love, hooks 2000.

honed in the anti-image of the nationalist, authoritarian and racist right, it often proceeds on the same epistemological basis.

Critical scholarship is routine fare in most left-leaning humanities and social sciences disciplines. Those concerned with capitalism and class address how we are alienated from our labour, from each other and from the planet around us. Those concerned with racism and colonialism explore how race and Empire function to organize humans into hierarchies to the ends of domination. These approaches overlap and share many features. Both stand in critical relation to the dominant ideal figure at the centre. For those focusing on race and colonialism this is routinely the white, colonial figure around whom racialized notions of humanity, morality and progress violently turn. For those focusing on class, the bourgeois or enlightenment man does similar work in explaining the violence of property and capitalism. In both cases the constitutive outside of these centres is the racialized and working-class human which the critical scholar labours to free, to different degrees and at different levels of abstraction.

These forms of scholarship, in their various guises, have been the default mode of critical left analysis for almost a century, arguably reaching their zenith in the post-structuralisms and post-colonialisms of the 1970s to 1990s. While post-structural and post-colonial analysis was problematic for scholar activists keen to uphold more clearly delineated political positions, for example on racism (Sivanandan 1990), it was also the case that post-structuralism and post-colonialism developed increasingly refined (crafted) languages and frameworks through which questions of racial and classed violence, Empire, alienation, subjectivity and power could be understood. This complexity was partly for its own sake, to be sure, but it also responded to the dynamic entanglement of social forces in late modern capitalist societies.

Post-2008 – 'the lost decade', the financial crisis, austerity, the killing of Mark Duggan, 'the refugee crisis', Prevent and the Hostile Environment – the confected hope of the new millennium feels eons away.[2] Pledges to end child poverty and arguments

---

[2] For detail, see Ali and Whitham 2018; Bakkali 2019; Bhattacharyya 2017; de Noronha 2019; El-Enany 2020; James 2014a; 2019; Jones et al. 2017; Kapoor 2018; Kundnani 2014; Rashid 2016; Trilling 2018; Valluvan et al. 2013.

over redistribution and the constitution of multi-ethnic Britain are now faint memories in a miasma of climate denial, fake news and nationalist-authoritarian assertion. And it is here that a more assertive critical left scholarship has emerged. In schools formerly entrenched in the post-structural analyses of Stuart Hall, Avtar Brah and others, new fractures have developed. Stuart Hall's death in 2014 in some way marks a watershed as scholars (often identifying as activists) moved away from the frameworks and idioms of post-structuralism towards particular readings of anti-colonial and anti-racist texts. Here, standards of colonial and anti-racist theory are revisited, separated out and pared down to fit the starker positions the moment necessitates. Fanon became a version of *Wretched of the Earth* (Fanon 1991) and Baldwin a partial distillation of *The Fire Next Time* (Baldwin 1973).

To say these analyses are formed on the same terrain as dominant culture is not to suggest they push in the same direction. The opposite is, of course, true. After all, critical traditions work to debunk racial and classed hierarchies and to call out its myths. Scholars inclined in this way take the categories and power structures employed by dominant versions of racial and classed modernity and show how they are bankrupt, how they are historically routed and how they act in the present. But such critiques map over the lines laid down by racial modernity itself. Their critique is maintained by positioning themselves in negative relation to the dominant movement – sometimes referring to themselves as 'resistance' on those terms. That is to say, increasingly strident and reactionary forms of nationalism and racism are matched up by more assertive anti-racist and anti-capitalist positions, but the terms of engagement are still set by racist and capitalist violence, as opposed to a world view that springs organically from modernity's counter-currents.

To develop this point, consider for a moment a different form of leftist critique against which left anti-racism is set. This is the left critique of capitalism that also frames the working-class as white (normally written 'white working-class'). Here the working-class is acknowledged as being constituted on the outside of bourgeois capitalism, and a particular reading of that marginal position is advanced in relation to neo-liberal capitalism and austerity. This version of critique calls out the class dimensions of social violence by mobilizing a standard working-class vs bourgeois capitalism argument, but does so, crucially, on the terms of whiteness. The figuration of whiteness is not idle. It turns the orthodox working-class vs bourgeois capitalism critique into an issue of race without developing a critique of racism. That is to say, the figuration of the working-class as white develops a critique of socio-economic injustice but does not extend the same critical lens to racism. Whiteness then operates as a reactionary category not a critical one. Unlike class, whiteness is not conveyed in critical relation to power but rather reactively solicits a key mode of human domination – namely, white supremacy. That claim is defensive and identitarian. Its identitarian defence turns around class loss, explained through the loss experienced at the hands of multi-ethnic society and non-white, or not white enough (in the case of Eastern Europeans) immigration (James 2014b; James and Valluvan 2018). Whiteness here is valorized as Belonging, the Nation and Right to its hard to win resources. Left scholarship of the white working-class is then less a defence of class against capital than it is a reactionary identitarian claim against those who are less racially deserving (Valluvan 2019). The identitarian claim to whiteness, and not the class critique, is the heart of the matter, and that explains the manifest inconsistencies of the argument. In the working-class-as-white approach, whiteness is not an anodyne supplement to class analysis but its driving force. The appeal to whiteness then relegates a critical analysis of class to little more than intellectual dressing for a reactionary racist claim.

Anti-racist scholarship routinely calls out this version of left critique for the assertion of racist dominance it so manifestly is. In so doing, it draws connections between the working-class-as-white argument and more assertive forms of white supremacy, nationalism and authoritarianism operating across the West and Europe. But while it debunks those myths, locating them in their wider historical projects of domination, it adopts similar epistemic terms. That is to say, its critique of racism is also sometimes based on defining tighter and more authentic identitarian claims against racist violence, often on the terrain established by racial dominance itself. This move has different names. It is sometimes called 'strategic essentialism', but it is also named the 'treacherous bind' (Spivak 2008: 260; Gunaratnam 2003: 28–9). Strategic essentialism lies in defining and defending people's history, culture and cosmologies against colonial and racist erasure, and in mounting resistance on the basis of those resources. Here structure, power, identity and becoming can be held productively in tension. The treacherous bind is a warning of how that identitarian strategy can slip from a position of critical defence and action to one in which essentialist orders set by racism and colonialism become the governing orders for minoritized groups – the kind of resistance that the 'nation-state hegemony finds logically comprehensible and ontologically satisfying' (Valluvan 2019: 47). When the strategic tensions between structure, power, identity, becoming and action slip into a starker defence of identitarian positions, those positions become the end in themselves, not a 'strategic' means to critical resistance. The defensive positions that reject racist austerity capitalism then acquiesce to the terms set by it.

Social media intensifies this landscape in particular ways, stripping anti-racist utterances from their social context and opening them onto the wider trends of hegemonic media flows. In Britain, US political positionalities become increasingly influential and accordingly regiment the tone of anti-racist debate. In so doing, the nuances and porosities of a more grounded dialogue are distilled and hyped through reactive Twitter contests (Seymour 2019; Titley 2019). There, affective claims and counter claims are won through amplifying sedimented positions, shored up by the mobilization of cultish celebrity. Our transient attentions are continually captured by arresting demonstrations of grievance and injustice.

There, the impression of collective anti-racist assertion is little more than overinvested digital personas. Such political action no longer has the means or desire to unravel its object of attention – racism. It no longer has the ability to form critical anti-racist associations. A politics that claims to do anti-racist work through loud, cultish and denuded accounts of elsewhere is not effectively anti-racist at all. It feeds a theatre of indignant contest set on its own erroneous terms.

Through these submissions we adhere to one of bourgeois capitalism's principle ruses, and operations of power – the ruse/power over how time is ordained to the end of property and profit. Because what is occurring in these moves is also an over-privileging of the immediate, in which the past becomes a fragment in critical scholarship to more forcefully render the anti-racist claim. Snippets of texts, bites of convenient history and the odd Lorde and Fanon quote become incorporated into a form of action less interested in establishing its relation to those pasts than it is in substantiating its own immediate claim through the currency of the present. As with all bourgeois capitalist regimes, the immediate must be dominated, must be owned, an assertion must be substantiated over that moment. Loudly contesting racial dominance while re-inscribing capitalist governance can then frustrate, rather than further, activist scholarship and its knowledge of located anti-racist histories and cultures.

Pushing back and forth along the lines set by modern dominance, this mode of critical left scholarship then finds it difficult to understand how cultural politics exist in alternative relation to dominant modernity. It finds it difficult to assess the ways in which alternative modernisms intertwine with dominance and how they are also available as powerful threads for anti-racist work. It finds it difficult not because it disagrees with the possibility of such things, but rather because the moulds in which it is sedimented frustrate those lines of interrogation and action. When scholarly engagement is set by the epistemic terms it contests, when its defensive identitarian positions pertain to the magnitudes of that which it disavows and when the effectivity of the argument in that arena comes through strategies of violence and property, it will be necessarily difficult both to understand the ambiguities of a moment and to have a feeling for the forms of critical modern life that move beneath the ground in which its feet are so firmly planted.

It then takes a riot, heavy snowfall (in the UK that can reroute society for a day), tragedies such as Grenfell and the 'refugee crisis' or a strike to bring popular attention to the cultural political alternatives of quotidian life.[3] When these disruptions take place, alternative conditions are awoken and, momentarily at least, frames of reference are shifted. When flows of commerce and drudgery are recomposed by fire and mass street occupation; when the middle-class and working-class sides of London's neighbourhoods convene for snowball fights and sledging; when hundreds walk silently every month against housing injustice; when strangers come together to care for refugees on shores of the Mediterranean; when workers act together against injustice, a reconnection with people and the deliberative reality of political action occurs. But as normal service resumes, those moments are folded back into the rhythms of dominance.

---

[3] Or, indeed, door-stepping in a General Election campaign (Balani 2019).

Critical left scholarship, then, offers an attractive thesis – one which can be mobilized with facility to understand all manner of social, cultural and political problems. Indeed, the most reductive and loudest versions of these critiques are particularly saleable and satisfying. When society is run by politicians who have more invested in hedge funds than high streets and where racism and nationalism are the principle modes of generating electoral consent, nothing feels better than to shout back, but a roar on those terms, however cathartic, is a cry in the night.

That is not the approach I have been endorsing in this book. Rather, through a discussion of sonic intimacy, the book has explored how versions of modern culture sit askew to the dominant regimes of racial capitalism – not with or without, but alternative; and how alternative culture provides political resources for anti-racist and post-colonial work.

Making this case for alternative political culture is categorically not an argument for the pre-modern – for animism and the like. I have repeatedly stated the primordial perils of these discussions in relation to race. While animist analyses might sometimes be useful in figuring industrial modernity as a destructive aberration in planetary time, their anti-modern positions also frustrate alternative engagements with modernity and the uneven resources it contains, both intellectual and political, for anti-racist and post-colonial scholarship. This book's analysis of sonic intimacy has responded to the racial and classed violence of late modernity not by addressing culture outside modernity (the primordial model) or culture as abject from modernity (resistance), but by addressing culture in alternative relation to dominant modernity.

In this book, the alternative cultural politics of black diasporic sound cultures have been discussed in terms of sonic intimacy. It has been through sonic intimacy that the book has reflected on the ways in which the cultural politics of the black Atlantic are alternative to dominant racial, capitalist visual and textual scripts. It is through the transformation of sonic intimacy that key shifts in alternative cultural politics have been assessed.

In this book, alternative cultural politics has been placed in tension with the dominant politics of racial capitalism – economist thinking, property ownership, individualism, racism, ethnocentricity and nationalism. Rather than focus on the sour words of politicians and the barbarous acts of the police and military, domination has been explored through popular culture-because it is in popular culture, in the fabric of everyday life and in the creative and habitual patternings of the ordinary, that barbarity is lived, and consent to it secured (Hall 1996). Equally, it is in popular culture that alternatives to barbarousness are found. It is in the routine banality of popular culture that lie vital contradictions and persistent sources of hope.

Sonic intimacy is suited to this inquiry precisely because of its ambivalent relationship to dominance – to visual and textual racial capitalism. The sonic intimacies of black diasporic culture then carry a cultural politics from below; a cultural politics which has other stories to tell. 'Below', here, does not denote 'separate' from dominant modernity. Rather, below is the minor key to modernity's major movements.[4] 'Below' is laid down against the grain of capitalist information-time. 'Below' prefigures dominant modern epistemes as it also carries the terror and joy of their times. What is 'below' can be struck up and rearranged for the present.[5]

'Below' is demotic, realized by revellers and open to fellow travellers. In diasporic form, it moves with and through time, across nations and racial order. Through citation it traces the violence of passage, and the spread of ecumenical knowledge. Across regimes of property and capital it flows half-articulate, half-hidden, claimed but not captured. 'Below', then, are the registers through which humanity might flourish. 'Below' are the voices, frequencies and vibrations of many alternative generations.

'Below' we depart from the comfortable terrains of claim and counter. And the impetus to do so is pressing. In this moment of heightened racist and authoritarian assertion 'pathways to dwell on what it means to be human' and vehicles through which 'we might give humanness a different form' matter deeply (McKittrick 2015: 10).

---

[4]This is an engagement with notions of 'below' and the demotic sketched in other post-colonial and critical race texts (see McKittrick 2015; McKittrick 2006).
[5]See, for historical discussion, Bakhtin 1986; Benjamin 1968.

Frequency

The discussion of alternative cultural politics in this book has been convoluted. It has explored shifts in mediation, and in the qualities and frequencies of sound. It has been concerned with alternative realizations of time and ways of knowing. It has considered shifting expressive relationships and mutualities. And it has addressed how all of these are held and conveyed through techno-social-sonic moments.

The journey from the sonic intimacies of dancehalls and radios to the screens of YouTube music videos offers one path for reflection. That path concerns the relationship between frequency and alternative cultural politics.

The reggae sound system's bass carried black diasporic wisdoms and depths – 'the moving hurting black story' – which travelled beyond the racial capitalism of the day. Jungle raves fired up that low-end strut through in-concert antiphonies to Yuppie life. Pirate radio quickened their movement and diminished their depths. Those rolling, accelerating, waning subs were intense. They are going to take over ... and then digital capitalism catches up. Grime YouTube music videos are lighter and quicker still. As digitized hype was habitually lived, bass was too heavy and the social and historical messages of pleasure and pain it carried gave way to featherweight sonics at historical odds with the alter-gravities of modernity.

But with grime's mobilities the voice of the MC also returned, displacing the DJ and his/her turntables from centre stage. MC vocals thrived in the treble range. A load lessened of vinyl, decks and weighty speaker stacks. Leaking out of East London's former industrial landscapes were tinny tonalities of humanity, not only white capitalist mobility. The MC's voice carried pathos and pleasure like no other instrument, and that resonated word for word with the neo-liberal city.

YouTube music videos altered those frequencies but the visual did not displace the sonic. Rather, social media screens start to sing. Sound was screened, and then the screen was sound; the humanism of the MC's voice re-coloured in visual digital flows. Snapshots of authenticity, allure and danger infiltrated the located complexity of sonic productions. Signifiers of racism, masculinity and personal brand patterned their phonics. Vocalizations of anger, warmth and coldness were coded onto the body. But the singing continued. The sound signed through the racialized body, sang with a human voice, a voice that appeared in pixelated form. And while that singing continues, racial capitalism has not taken full hold.

Autonomous temporal zones

Travels in time, from clockwise to 'dub-wise', the alternative temporalities of the reggae sound system move outside the grind of white bourgeois rhythmatics, keeping alive black diasporic experiences at home and abroad. Jungle pirate radio cites the chronometers of dub inducing autonomous temporal zones of intensity and speed. Dance floors and call-ins strain the mechanisms of the normative dial, the aesthetic impermanence of dialogue and mixing distort diurnal formality, for the reimagination of the future city.

Grime is more immediate still; on the limit, on the brink, spilling out of pirate radio studios across London, rooftops with no safety rail. Its raves and radios move in lockstep with the grit and impermanence of advancing marginality. Raves, record shops, radio ... YouTube music videos intensify that immediacy, but also curtail its impermanence. On impermanence, post-reggae humanisms were set, and through YouTube they decay. In flux but not on the brink, grime YouTube music videos are out-of-time hollow hyper replications of the now, intensifying grime's theatre of claim and counter-claim to the detriment of urgently restlessness and autonomous temporality. YouTube grime music videos move with the compulsive clicks of the networked meter, when breaking the mould of dominant time still matters a great deal.

Wisdom

The knowledge systems of reggae, jungle and grime move outside the property regimes of dominant life. Livity, junglism and grime lay below the informatic-now. There the wisdoms of black life in British cities are compiled with acumens of the diasporic experience, intelligence on fast-paced multi-ethnic conviviality and the aptitude of post-industrial places. In series, they are chains of activation. Together they carry the fullness of shared life, of knowing touched by many hands.

Wisdom moves through music – through codified sounds played back to the city and affirmed by its celebrants. Affirmations of social position and daily rituals, political struggle, oppression and emancipation filter in and out of those sounds. But wisdom is borne through sonic atmospherics, too. Registers of intensity, coldness and warmth, structure the general knowledge of moments, as their variable amplification patterns sagacity in time. Intertwined with information, in reggae and jungle wisdom nonetheless comes first. Darkness, versioning, white labels, mixing and the sonic form itself undermine property. But in grime YouTube music videos, wisdom and information are more tightly twisted. Dominance is nearly indiscernible from its alternatives, although alternatives persist nonetheless. In fact, DIY grime YouTube productions are redolent with alternative knowledge, even as they are routed through the reductive data streams of digital capitalism.

Craft

Craft doesn't live here any more! Those techno-social processes anathema to labour alienation, those open material relations productively at odds with capitalist exploitation, seem to have had their day with the advent of YouTube and 3G. In the reggae sound system, constellations of skill, refinement and creativity found their expressive outlets through the sounds and sociabilities of the dance floor. Consumer culture made inroads into jungle but couldn't pierce the illegality of its pirate infrastructure where apprenticeship thrived. As the tower blocks fell and the social media came of age, the closed circuitry of consumer technology advanced. That didn't change the music or its feeling overnight, but its techno-social relations were altered. Craft has been laid down, and with it a potent alternative to techno-capitalist domination.

Mutuality

Mutuality has transformed. In reggae sound systems and jungle pirate radio, mutuality was sustained through finitude – presence and horizontal relations within delimited social and geographical spaces. Finitude facilitated the openness on which the more routine appraisal of multi-ethnic conviviality is based (Back 1994; Gilroy 2004; James 2015; Valluvan 2016). Finitude structures a city's 'indifference to difference' (Amin 2010). The finite collective of the dancehall and the pirate radio made contact and coexistence comprehensible.

YouTube is plenitude. Grime YouTube music videos are still convivial, but their mediation is not finite. YouTube music videos require users to infill the absence of the everywhere-but-nowhere interlocutor. That absence is often supplemented by retrenchment to the self and loud saleable identity positions. In that sense, plenitude is the feeling of being pulled apart. It is the basis of anxiety which feeds the already existing social negations of late modern life.

Mutuality doesn't end there. It also encompasses the politics of care and gift that are inherent to conviviality but do not require it and cannot be reduced to it. Mutuality, care and gifting operate as alternatives to cruelty, property ownership and privatization, even as they are often entangled with them. In reggae, jungle and grime, they are observed through community, kinship and the intergenerational spread of family networks and peer associations. These relations are deep. They work against the privatization dominant society demands. They extend from close networks to wider communities, to the 'massive', where they are regarded as 'life support systems'. An extraordinary claim in today's terms, the systems of care in sound cultures can support life!

And in sonic life-support systems, expression and dialogue routinely come before profit. Here, wisdom is gifted through apprenticeship, music is gifted to the city, producers gift beats, mixes and samples to one another. Property is not the point. Getting rich or getting by is usually secondary. And even when such practices find affinity with prosumption, they are not totally captured by it. Reciprocation is still vital. And in the context in which hostility is the norm, that matters also.

Wholeness

Together, these alternative cultural politics are often explained as magic and mysticism. This is not because ravers believe in witches and steppas in wizards, but because the experience of sound cultures surpasses language. But this experience isn't magic, it's better than magic. It's the full and relational coming together of diasporic sounds, technologies and human/social relations in the sound culture as expressed in the dancehall, rave or radio. That particular collection of factors coheres into a sense of something powerful, something genuinely spellbinding.

The vibe of the reggae dancehall is a wholeness of haptics and low-end connectivity. It is a wholeness that precedes the moment that the stylus touches the record. As the tune drops, technology, sound and the social are activated. The wholeness of jungle pirate radio is speed and love. It is raves, radios, cars, phones and records, running to burn-out. That 'hype' is mutual, outward-looking, multi-ethnic, black and expansive, but within the city's limits. It is composed of multiple diasporic strands. Hype is for/by the junglist massive. It is prefigured in the ether. It anticipates the MC. It hangs for the b-line, drop, buzzing for the phone-in, for the shout to Anita from Charlton. Full to bursting, jungle pirate radio wholeness is supercharged. It pretends to obliterate the moment, but it can't.

The wholeness of grime YouTube music videos is viscosity, endorphins, blue light, paper-thin permanence and hyperlinked bits. It is social media, mobile phones, tinny speakers, screens, the Internet and the everywhere/nowhere digital capitalist prosumer. The affective binds, chemical rushes and restless hyperactivity provide feelings of hunger and near-satiation, a sense of spilling over that is too capital-ordained to ever really be poetic. Its spills and thrills are less based on a dissonant fullness at odds with dominant racial capitalism than they are of the denial of that condition. This is the wholeness of the lived multi-ethnic city, for a sound culture that is not easily ethnically owned. But it is also a wholeness for audiences sourced by Google Analytics, for spectators fed on hackneyed images of authentic saleable selves.

The sound cultures discussed in this book do not exhaust the sonic intimacies of late modernity, or even those of black diasporic cultures, but they are three of the most popular and most telling. Shifts between them matter because they concern our abilities to be filled with the other, with the dance, with the city and with humanness. Vibes, hypes and grimes instruct us on the resonance and residue of humanity. Their challenge is where to press one's ear and how to listen. In an era better attuned to vociferous claim and counter claim, the minor keys of human life easily go unnoticed, when they have much to offer. The sonic intimacy of black diasporic sound cultures is an avenue through which these alternatives can be heard. Relation, craft, great time, wisdom, finitude, mutuality and wholeness exist as possible pathways to greater mutual living. It is in their autonomy and poetry that humanity could thrive, and it has been in the *sonic* intimacy of black diasporic culture, at least until now, that (some) of their vitality has been sustained.

REFERENCES

24H Canal +. 1994. 'Jungle and Drum & Bass in London'. Available online: https://www.youtube.com/watch?v=SghWOAjiVYU (accessed 23 April 2018).

A Plus. 'Conflict DVD'. Available online: https://www.youtube.com/watch?v=qqqwq9mxUPc (accessed 20 January 2020).

Airoldia, Massimo, et al. 2016. 'Follow the Algorithm: An Exploratory Investigation of Music on YouTube'. *Poetics* no. 57: 1–13.

Alexander, Claire. 1996. *The Art of Being Black: The Creation of Black British Youth Identities*. Oxford: Clarendon Press.

Ali, Nadya and Ben Whitham. 2018. 'The Unbearable Anxiety of Being: Ideological Fantasies of British Muslims Beyond the Politics of Security'. *Security Dialogue*, 49, no. 5: 400–17.

Amin, Ash. 2010. 'The Remainders of Race'. *Theory, Culture and Society*, 27, no. 1: 1–23.

Arendt, Hannah. 1958. *The Human Condition*. 2nd edn [reissued with improved index and new introduction by Margaret Canovan]. Chicago; London: University of Chicago Press, 1998.

Arendt, Hannah. 1978. *The Life of the Mind*. New York; London: Harcourt Brace Jovanovich.

Arte Vost. n.d. 'Jah Shaka 90'. Available online: https://www.youtube.com/watch?v=yAIBuui6mLY (accessed 15 July 2019).

Back, Les. 1988. 'Coughing Up Fire: Sound Systems in South-East London'. *New Formations*, 5 (summer): 141–52.

Back, Les. 1994. *New Ethnicities and Urban Culture: Racisms and Multiculture in Young Lives*. London: UCL Press.

Back, Les. 2003. 'Sounds in the Crowd'. In Michael Bull and Les Back (eds), *The Auditory Culture Reader*, pp. 311–27. Oxford: Berg.

Bakhtin, Mikhail. 1986. 'Toward a Methodology of the Human Sciences'. In M. M. Bakhtin, Caryl Emerson and Michael Holquist (eds), *Speech Genres and Other Late Essays*, pp. 157–70. Austin: University of Texas Press.

Bakkali, Yusef. 2019. 'Dying to Live: Youth Violence and the Munpain'. *The Sociological Review*, 67, no. 6: 1317–32.

Balani, Sita. 2019. 'Dangerous Spaces'. Available online: https://www.versobooks.com/blogs/4518-dangerous-spaces (accessed 20 January 2020).

Baldwin, James. 1973. *The Fire Next Time*. London: Penguin.

Bass Culture Research. 2018. 'Blacker and Molly Full Interview'. Available online: http://basscultureduk.com/blacker-and-molly-full-interview/ (accessed 15 July 2019).

Bauman, Zygmunt. 1998. *Work, Consumerism and the New Poor*. Buckingham: Open University Press.

Bauman, Zygmunt. 2000. *Liquid Modernity*. Cambridge: Polity Press.

BBC Radio 1Xtra. 2015. 'Big Narstie Keeps it Real About Grime'. Available online: https://www.youtube.com/watch?v=kEwv8xOLUI0 (accessed 20 January 2020).

BBC. 2019. 'YouTube Is Most-Watched Platform for Young People, Says Report'. Available online: https://www.bbc.co.uk/newsround/49261194 (accessed 20 January 2020).

BBC2. 1994. 'Arena Radio Night'. Available online: https://www.youtube.com/watch?v=nbQU5UaMf20 (accessed 24 April 2018).

Beauvoir, Simone de and Gisèle Halimi. 1962. *Djamila Boupacha: The Story of the Torture of a Young Algerian Girl which Shocked Liberal French Opinion*. London: André Deutsch and George Weidenfeld and Nicolson.

Beauvoir, Simone de. 2009 [1972]. *The Second Sex*. New edn, trans. Constance Borde and Sheil Malovany-Chevallier. London: Jonathan Cape.

Beck, Ulrich and Elisabeth Beck-Gernsheim. 2001. *Individualization: Institutionalized Individualism and its Social and Political Consequences*. London: SAGE.

Belle-Fortune, Brian. 2004. *All Crews: Journeys Through Jungle*. London: Vision Publishing.

Benjamin, Walter and Hannah Arendt. 1968. *Illuminations*. London: Fontana, 1992.

Benjamin, Walter. 1968. 'Thesis on the Philosophy of History'. In Walter Benjamin (ed.), *Illuminations*, pp. 253–64. New York: Schocken Books.

Benjamin, Walter. 1979. 'Naples'. In Walter Benjamin (ed.), *One-Way Street, and Other Writings*, pp. 167–76. London: NLB.

Berland, Jody. 1993. 'Sound Image and Social Space: Music Video and Media Reconstruction'. In Simon Frith, Andrew Goodwin and Lawrence Grossberg (eds), *Sound and Vision: The Music Video Reader*, pp. 20–36. London: Routledge.

Berlant, Lauren Gail. 2008. *The Female Complaint: The Unfinished Business of Sentimentality in American Culture*. Durham, NC: Duke University Press.

Berlant, Lauren Gail. 2011. *Cruel Optimism*. Durham, NC: Duke University Press.
Berlant, Lauren. 1998. 'Intimacy: A Special Issue'. *Critical Inquiry*, 24, no. 2: 281–8.
Bhattacharyya, Gargi. 2017. *Rethinking Racial Capitalism: Questions of Reproduction and Survival*. London: Rowman and Littlefield.
Bizzle, Lethal. 2006. 'David Cameron is a Donut'. Available online: https://www.theguardian.com/commentisfree/2006/jun/08/davidcameronisadonut (accessed 20 January 2020).
Boutin, Aimee. 2015. *City of Noise: Sound and Nineteenth-Century Paris*. Chicago: University of Illinois Press.
Bovell, Dennis and Les Back. 2017. 'Dub on Air with Dennis Bovell'. Available online: https://www.mixcloud.com/sohoradio/dub-on-air-with-dennis-bovell-22012017/ (accessed 4 January 2020).
Boym, Svetlana. 2001. *The Future of Nostalgia*. New York: Basic Books.
Bradley, Lloyd. 2000. *Bass Culture: When Reggae was King*. London: Viking.
Bramwell, Richard. 2015. *UK Hip-Hop, Grime and the City: The Aesthetics and Ethics of London's Rap Scenes*. London: Routledge.
Bull, Michael. 2000. Sounding out the city: personal stereos and the management of everyday life. Oxford: Berg.
Callahan, John F. 1996. 'Introduction'. In Ralph Ellison and John F. Callahan (eds), *Flying Home and Other Stories*, pp. ix–xxxviii. New York: Random House.
Campion, Karis. 2017. 'Making Mixed Race: Time, Place and Identities in Birmingham'. PhD thesis, University of Manchester.
Carby, Hazel V. 1987. *Reconstructing Womanhood: The Emergence of the Afro-American Woman Novelist*. Oxford: Oxford University Press.
Chambers, Iain. 1985. *Urban Rhythms: Pop Music and Popular Culture*. London: Macmillan.
Charles, Monique. 2018. 'MDA as a Research Method of Generic Musical Analysis for the Social Sciences: Sifting Through Grime (Music) as an SFT Case Study'. *International Journal of Qualitative Methods*, 17: 1–11.
Christoloudou, Chris. 2009. 'Renegade Hardware: Speed, Pleasure and Cultural Practice in Drum 'n' Bass'. Thesis, South Bank University.
Clifton, Jamie. 2015. 'Jungle, Raves and Pirate Radio: The History and Future of Kool FM'. Available online: https://www.vice.com/en_uk/article/4wba59/kool-fm-jungle-pirate-london-945 (accessed 20 January 2020).
Collin, Matthew. 1997. *Altered State: The Story of Ecstasy Culture and Acid House*. London: Serpent's Tail.
Collins, Hattie and Olivia Rose. 2016. *This is Grime*. London: Hodder and Stoughton.
Cordell, Tom. 1997a. 'Kool FM'. Available online: https://vimeo.com/26938088 (accessed 20 January 2020).

Cordell, Tom. 1997b. 'One Nation NYE 1997'. Available online: https://vimeo.com/267701915 (accessed 20 January 2020).
Cordell, Tom. 1997c. 'Rave Cliptape'. Available online: https://vimeo.com/164007089/4d416e12dc (accessed 20 January 2020).
Coxsone, Lloyd. 2018. 'The Beginning of my Career in Sound'. Available online: http://basscultureduk.com/sir-lloyd-coxsone-the-beginning-of-my-career-in-sound/ (accessed 17 July 2019).
Crisell, Andrew. 1994. *Understanding Radio*. 2nd edn. London: Routledge.
Davidson, James, et al. 2010. 'The YouTube Video Recommendation System'. RecSys2010. Barcelona, Spain.
de Noronha, Luke. 2019. 'Deportation, Racism and Multi-Status Britain: Immigration Control and the Production of Race in the Present'. *Ethnic and Racial Studies*, 42, no. 14: 2413–30.
Dean, Jodi. 2005. 'Communicative Capitalism: Circulation and the Foreclosure of Politics'. *Cultural Politics*, 1, no. 1: 51–74.
Dean, Jodi. 2010. 'Affective Networks'. *Media Tropes*, 2, no. 2: 19–44.
Debord, Guy. 2002. *Society of the Spectacle*. London: Rebel Press.
Deleuze, Gilles and Felix Guattari. 2004. *A Thousand Plateaus: Capitalism and Schizophrenia*. London: Continuum.
Deleuze, Gilles. 1987. *Dialogues*. London: Athlone.
Deleuze, Gilles. 1990. *Expressionism in Philosophy: Spinoza*. New York: Zone Books.
Deleuze, Gilles. 1997. *Essays Critical and Clinical*. Minneapolis: University of Minnesota Press.
DeNora, Tia. 1999. 'Music as a Technology of the Self'. *Poetics*, 27: 31–56.
DeNora, Tia. 2000. *Music in Everyday Life*. Cambridge: Cambridge University Press.
Dineen, Molly and Julian Caidan. 1981. 'Sound Business'. Available online: https://www.youtube.com/watch?v=0xuObz5d6BI (accessed 15 July 2019).
Dizzee Rascal and Slimzee. c. 2001. 'Dizzee Rascal & Slimzee at Sidewinder Milton Keynes'. Available online: https://www.youtube.com/watch?v=z9RxqXZ9bxk (accessed 10 February 2020).
DJ Brockie and MC Det. 1996. Kool FM.
DJ Sparkie. 1994. Freedom FM 86.6.
DJ Target. 2018. *Grime Kids*. London: Trapeze.
DJ Wicked et al. 1992. Defection FM 89.4.
DJ Zinc and MC Rage. 1995. Eruption 101.3.
DJDUBLTV. 2017. 'Devlin Interview'. Available online: https://www.youtube.com/watch?v=RFDvyIao0Ys (accessed 20 January 2020).
Doran, John. 2014. 'Radio Live Transmission: 22 Years of Pirate Broadcasts with Rude FM'. Available online: http://thequietus.com/articles/14317-rude-fm-jungle-hardcore-pirate-radio (accessed 27 February 2018).

Douglas, Susan J. 2004. *Listening In: Radio and the American Imagination*. Minneapolis, MN: University of Minnesota Press.
Douglass, Frederick. 2001. *Narrative of the Life of Frederick Douglass: An American Slave*. New Haven: Yale University Press.
Dread, Mikey. 2012. 'Channel One Sound System – Red Bull Culture Clash – Interview'. Available online: https://www.youtube.com/watch?v=cesi52_MfN4 (accessed 17 July 2019).
Du Bois, W. E. B. 2007. *The Souls of Black Folk*. Oxford: Oxford University Press.
Dub, Upsetters 14. 1973. *Black Board Jungle*. Upsetter.
Dugs, Uncle and Andrew Woods. 2016. *Rave Diaries and Tower Block Tales*. London: Music Mondays.
Dyer, Richard. 1997. *White*. London: Routledge.
Dyson, Frances. 1994. 'The Genealogy of the Radio Voice'. In Daina Augaitis and Dan Lander (eds), *Radio Rethink: Art, Sound, and Transmission*, pp. 167–88. Banff, Alberta: Walter Phillips Gallery.
El-Enany, Nadine. 2020. *(B)ordering Britain: Law, Race and Empire*. 1st edn. Manchester: University of Manchester Press.
Ellington, Duke. 1993. *The Duke Ellington Reader*. Oxford: Oxford University Press.
Ellison, Ralph. 1995. 'Living with Music'. In John F. Callahan (ed.), *The Collected Essays of Ralph Ellison*, pp. 227–63. New York: Modern Library.
Ellison, Ralph. 2001. *Invisible Man*. London: Penguin.
Eshun, Kodwo. 1998. *More Brilliant than the Sun: Adventures in Sonic Fiction*. London: Quartet Books.
Evans, Ollie. 2014. 'Music Nation – Jungle Fever'. UK. Available online: https://vimeo.com/104712313 (accessed 29 April 2020).
Fanon, Frantz. 1986. *Black Skin, White Masks*. London: Vintage.
Fanon, Frantz. 1991. *The Wretched of the Earth*. New York: Grove Weidenfeld.
Ferrigno, Emily Daus. 2008. 'Technologies of Emotion: Creating and Performing Drum 'n' Bass'. Middletown, CN: Wesleyan University.
Fisher, Martin and George Stricker. 1982. 'Preface'. In Martin Fisher and George Stricker (eds), *Intimacy*, pp. xi–xii. New York: Plenum.
Folke, Karl and Andreas Weslien. 2008. 'Musically Mad – UK Sound System Documentary'. Available online: https://www.youtube.com/watch?v=NK0vBKsK7fI (accessed 15 July 2019).
France, Alan. 2007. *Understanding Youth in Late Modernity*. Maidenhead: Open University Press.
Fuchs, Christian. 2014. 'Digital Prosumption Labour on Social Media in the Context of the Capitalist Regime of Time'. *Time & Society*, 23, no. 1: 97–123.

Fuller, Matthew. 2005. *Media Ecologies: Materialist Energies in Art and Technoculture*. Cambridge, MA: MIT.

Gannes, Liz. 2009. 'YouTube Doubles Upload Size Limit'. Available online: https://gigaom.com/2009/06/26/youtube-doubles-upload-size-limit/ (accessed 28 January 2015).

Gaonach, Alexandre. 2014. 'United We Stand'. Available online: https://www.youtube.com/watch?v=OK3DA6cVrjo (accessed 15 July 2018).

Garland, Emma. 2017. 'NoLay is a Get-Shit-Done Force to be Reckoned With'. Available online: https://www.vice.com/en_uk/article/vv57qd/nolay-interview-2017-activism-this-woman (accessed 6 January 2020).

Gayle, Carl. 1974. 'The Reggae Underground'. *Black Music*, 1, no. 8: 14–20.

Gibbins, Paul. 2015. 'Lock Down Your Aerial'. Available online: https://mixmag.net/read/lock-down-your-aerial-blog (accessed 20 January 2020).

Giddens, Anthony. 1992. *The Transformation of Intimacy: Sexuality, Love and Eroticism in Modern Societies*. Cambridge: Polity.

Giggs. 2009. '"Talking the hardest": YouTube/UKgrimeTV'. Available online: http://www.youtube.com/watch?v=UpqQxklRkHU (accessed 15 August 2012).

Gilbert, Jeremy. 2004. 'Signifying Nothing: "culture", "discourse" and the Sociality of Affect', vol. 6. Available online: http://www.culturemachine.net/index.php/cm/article/view/8/7 (accessed 19 July 2016).

Gilroy, Paul. 1987. *'There ain't no black in the Union Jack': The Cultural Politics of Race and Nation*. London: Hutchinson.

Gilroy, Paul. 1993. *The Black Atlantic: Modernity and Double Consciousness*. London: Verso.

Gilroy, Paul. 2002 [1987]. *There Ain't No Black in the Union Jack: The Cultural Politics of Race and Nation*. London: Routledge.

Gilroy, Paul. 2004. *After Empire: Melancholia or Convivial Culture?* London: Routledge.

Glissant, Edouard. 1997. *Poetics of Relation*. Ann Arbor: University of Michigan Press.

Goddard, Grant. 2011. *Kiss FM: From Radical Radio to Big Business the Inside Story of a London Pirate Radio Station's Path to Success*. London: Radio Books.

Goodman, Steve. 2007. 'Contagious Transmission: On the Virology of the Pirate Radio'. In Erik Granly Jensen and Brandon LaBelle (eds), *Radio Territories*, pp. 49–54. Los Angeles: Errant Bodies Press.

Goodman, Steve. 2009. *Sonic Warfare: Sound, Affect, and the Ecology of Fear*. Cambridge, MA; London: MIT Press.

Grossberg, Lawrence. 1991. 'Rock, Territorialization and Power'. *Cultural Studies*, 5, no. 3: 358–67.

Guattari, Felix. 1995. *Chaosmosis: An Ethico-Aesthetic Paradigm*. Bloomington, IN: Indiana University Press.
Gunaratnam, Yasmin. 2003. *Researching Race and Ethnicity: Methods, Knowledge and Power*. London: SAGE.
Gutzmore, Cecil. 1993. 'Carnival, the State and the Black Masses in Britain'. In Winston James and Clive Harris (eds), *Inside Babylon: The Caribbean Diaspora in Britain*, 207–30. London: Verso.
Habermas, Jürgen. 1989. *The Structural Transformation of the Public Sphere: An Inquiry into a Category of Bourgeois Society*. Cambridge: Polity.
Hall, Stuart, et al. 1978. *Policing the Crisis: Mugging, the State, and Law and Order*. Basingstoke: Macmillan.
Hall, Stuart. 1996. 'Race: The Floating Signifier'. Available online: http://video.google.com/videoplay?docid=-8471383580282907865 (accessed 25 March 2009).
Hall, Stuart. 1997. 'The Spectacle of the "other"'. In Stuart Hall (ed.), *Representation: Cultural Representations and Signifying Practices*, pp. 223–90. London: Sage in association with the Open University.
Hall, Stuart. 2009. 'The "West Indian" Front Room'. In Michael McMillan (ed.), *The Front Room: Migrant Aesthetics in the Home*, pp. 17–23. London: Black Dog Publishing.
Hancox, Dan. 2009. 'On the Buses: Sodcasting and Mobile Music Culture'. Available online: https://dan-hancox.blogspot.com/2009/10/on-buses-sodcasting-and-mobile-music.html (accessed 20 January 2020).
Hancox, Dan. 2011. 'Pirate Radio Rave Tapes: "You can't Google this stuff"'. Available online: https://www.theguardian.com/music/2011/sep/08/pirate-radio-rave-tapes (accessed 27 February 2018).
Hancox, Dan. 2013. *Stand Up Tall: Dizzee Rascal and the Birth of Grime*. Kindle.
Hancox, Dan. 2015a. 'How British MCs Found a Voice of their Own'. Available online: https://www.theguardian.com/music/2015/may/31/british-mcs-stormzy-jammz-little-simz-krept-konan-novelist (accessed 5 October 2018).
Hancox, Dan. 2015b. 'Skepta's Mission'. Available online: http://www.thefader.com/2015/06/04/skepta-cover-story-konnichiwa-interview (accessed 5 October 2018).
Hancox, Dan. 2018. *Inner City Pressure: The Story of Grime*. London: William Collins.
Hartley, Bryan. 1971. 'Lessons of the Metro'. *Race Today*, 3, no. 10: 346–7.
Hattenstone, Simon. 2010. 'Dizzee Rascal: Fight to the Top'. Available online: https://www.theguardian.com/music/2010/jul/31/dizzee-rascal-interview (accessed 23 November 2018).

Hebdige, Dick. 1979. *Subculture: The Meaning of Style*. London: Methuen.
Hebditch, Stephen. 2015. *London's Pirate Pioneers*. London: TX Publications.
Hebditch, Stephen. 2017. *Pirate Radio Dispatches from Eighties London*. London: TX Publications.
Henriques, Julian. 2008. 'Sonic Diaspora, Vibrations, and Rhythm: Thinking Through the Sounding of the Jamaican Dancehall Session'. *African and Black Diaspora: An International Journal*, 1, no. 2: 215–36.
Henriques, Julian. 2011. *Sonic Bodies: Reggae Sound Systems, Performance Techniques, and Ways of Knowing*. London: Continuum.
Henry, William. 2006. *What the Deejay Said: A Critique from the Street!* London: Nu-Beyond.
Hind, John and Stephen Mosco. 1985. *Rebel Radio: The Full Story of British Pirate Radio*. London: Pluto.
Hirsch, Shirin. 2018. *In the shadow of Enoch Powell: race, locality and resistance*. Manchester: Manchester University Press.
hooks, bell. 2000. *All About Love: New Visions*. New York: William Morrow.
Howe, Darcus. 1973. 'Fighting Back: West Indian Youth and the Police in Notting Hill'. *Race Today*, 5, no. 11: 333–4.
Huxtable, Paul. 2014. *Sound System Culture: Celebrating Huddersfield's Sound Systems*. London: One Love Books.
Hyponik. 2016. 'DJ Zinc: blessed'. Available online: https://hyponik.com/features/dj-zinc-blessed/ (accessed 20 January 2020).
Ice Films Entertainment, et al. 2011a. 'Kill All a Dem'. Mussa Abdalla, Ice Films Entertainment. Available online: http://www.youtube.com/watch?v=SsalihiYtds (accessed 16 August 2012).
Ice Films Entertainment, et al. 2011b. 'Time is Up'. Mussa Abdalla, Ice Films Entertainment. Available online: http://www.youtube.com/watch?v=mbe_YBObzcg (accessed 16 August 2012).
Illouz, Eva. 2012. *Why Love Hurts: A Sociological Explanation*. Cambridge: Polity.
Institute of Race Relations. 1987. *Policing Against Black People*. London: Institute of Race Relations.
Iton, Richard. 2008. *In Search of the Black Fantastic: Politics and Popular Culture in the Post-Civil Rights Era*. Oxford: Oxford University Press.
Jackson, Rollo. 2011. *Tape Crackers*. London: Rollo Jackson.
Jah Shaka. 2014. 'Jah Shaka Lecture'. Available online: https://www.youtube.com/watch?v=3QNWpnwWgc4 (accessed 15 July 2019).
James, Malcolm and Sivamohan Valluvan. 2018. 'Left Problems, Nationalism and the Crisis'. *Salvage*, 6: 165–76.
James, Malcolm. 2014a. 'Mark Duggan and Britain's Post-Colonial Politics of Death', no. 5. Available online: http://www.discoversociety.org/2014/02/15/mark-duggan-and-britains-post-colonial-politics-of-death/ (accessed 22 May 2014).

James, Malcolm. 2014b. 'Whiteness and Loss in Outer East London: Tracing the Collective Memories of Diaspora Space'. *Ethnic and Racial Studies*, 37, no. 4: 652–67.
James, Malcolm. 2015. *Urban Multiculture: Youth, Politics and Cultural Transformation in a Global City*. Basingstoke: Palgrave.
James, Malcolm. 2018. 'Authoritarian Populism | Populist Authoritarianism'. In Alberto Duman, Dan Hancox, Malcolm James and Anna Minton (eds), *Regeneration Songs: Sounds of Loss and Investment from East London*, pp. 291–307. London: Repeater.
James, Malcolm. 2019. 'Care and Cruelty in Chios: The "Refugee Crisis" and the Limits of Europe'. *Ethnic and Racial Studies*, 42, no. 14: 2470–89.
James, Martin. 1997. *State of Bass: Jungle - the Story So Far*. London: Boxtree.
Jameson, Fredric. 1991. *Postmodernism, or, the Cultural Logic of Late Capitalism*. London: Verso.
JamieS23. 2014. 'Reloading the Pirate Radio Dial'. Available online: https://breakbeat.co.uk/features/reloading-pirate-radio-dial/ (accessed 27 February 2018).
Jasen, Paul C. 2016. *Low End Theory: Bass, Bodies and the Materiality of Sonic Experience*. London: Bloomsbury.
JME for Noisey. 2014. 'The Police vs Grime Music – A Noisy Film'. Available online: https://www.youtube.com/watch?v=eW_iujPQpys (accessed 23 November 2018).
Johnson, Linton Kwesi. 1980. *Reggae Sounds*. Island Records.
Johnson, Linton Kwesi. 1981. '"Black Beard" in Profile: Laying the Foundations'. *Race Today Review*, 14, no. 1: 9–11.
Jones, Hannah, et al. 2017. *Go Home?: The Politics of Immigration Controversies*. Manchester: University of Manchester Press.
Jones, Simon. 1988. *Black Culture, White Youth: The Reggae Tradition from JA to UK*. Basingstoke: Macmillan Education.
Jordan, Tim. 2015. *Information Politics*. London: Pluto.
Kano. 2005a. *Nobody Don't Dance No More*. 679 Recordings.
Kano. 2005b. *Ps and Qs*. 679 Recordings.
Kaplan, E. Ann. 1987. *Rocking Around the Clock: Music Television, Postmodernism, and Consumer Culture*. London: Methuen.
Kapoor, Nisha. 2018. *Deport, Deprive, Extradite: 21st Century State Extremism*. London: Verso.
Kim, Helen. 2012. 'A "desi" Diaspora? The Production of "desiness" and London's Asian Urban Music Scene'. *Identities: Global Studies in Culture and Power*, 19, no. 5: 557–75.
Kim, Helen. 2015. *Making Diaspora in a Global City: South Asian Youth Cultures in London*. London: Routledge.

King Tubby and the Upsetter. 2005. *King Tubby Meets the Upsetter at the Grassroots of Dub*. Studio 16.

Kundnani, Arun. 2014. *The Muslims Are Coming!: Islamophobia, Extremism, and the Domestic War on Terror*. London: Verso.

La Rose, John. 1984. 'The New Cross Fire'. *Race Today: The Voice of the Black Community in Britain*, 15, no. 6: 3.

LaBelle, Brandon. 2010. *Acoustic Territories: Sound Culture and Everyday Life*. London: Continuum.

Lefebvre, Henri and Catherine Regulier. 2004. 'Attempt at the Rhythmanalysis of Mediterranean Cities'. In Henri Lefebvre (ed.), *Rhythmanalysis: Space, Time and Everyday Life*, pp. 85–100. London: Continuum.

Levi, Primo. 1988. *The Drowned and the Saved*. London: Penguin.

Levinas, Emmanuel. 1989. *The Levinas Reader*. Oxford: Basil Blackwell.

Levinas, Emmanuel. 1998. *Entre nous: Thinking-of-the-Other*. London: Continuum, 2006.

Lewis, Gail. 2012. 'Where Might I Find You? Objects and Internal Space for the Father'. *Psychoanalysis, Culture & Society*, 17, no. 2: 137–52.

Linke, Gabriele. 2011. 'The Public, the Private, and the Intimate: Richard Sennett's and Lauren Berlant's Cultural Criticism in Dialogue'. *Biography*, 34, no. 1: 11–24.

London Tonight. 1994. 'Report on Weekend Rush'. Available online: https://www.youtube.com/watch?v=mpS0jR6FG1o (accessed 24 April 2018).

Lorde, Audre. 1984. *Sister Outsider: Essays and Speeches*. Trumansburg, NY: Crossing Press.

Loviglio, Jason. 2005. *Radio's Intimate Public: Network Broadcasting and Mass-Mediated Democracy*. Minneapolis: University of Minnesota.

Lowe, Lisa. 2015. *The Intimacies of Four Continents*. Durham, NC: Duke University Press.

Marar, Ziyad. 2012. *Intimacy: Understanding the Subtle Power of Human Connection*. Durham, NC: Acumen.

Marcel, Gabriel. 1948. *The Philosophy of Existence*. London: Harvill Press.

Marley Marl and Footloose. 1995. Kool FM 94.5.

Marre, Jeremy. 2011. 'Reggae Britannia'. Available online: https://www.youtube.com/watch?v=wQXyK1uxTps (accessed 15 July 2019).

Marshall, Wayne. 2014. 'Treble Culture'. In Sumanth S. Gopinath and Jason Stanyek (eds), *The Oxford Handbook of Mobile Music Studies Vol. 2*, pp. 43–76. Oxford: Oxford University Press.

Massumi, Brian. 1995. 'The Autonomy of Affect'. *Cultural Critique*, 31, no. 2 (Autumn): 83–109.

Massumi, Brian. 2002. *Parables for the Virtual: Movement, Affect, Sensation*. Durham, NC: Duke University Press.
Mauss, Marcel. 1966. *The Gift: Forms and Functions of Exchange in Archaic Societies*. London: Cohen & West.
Mavros, Thalia. 2011. 'Palladium Presents: London Pirate Radio'. Available online: https://www.youtube.com/watch?v=ZpTJaTiD71Y (accessed 24 April 2018).
May, Chris. 1977a. 'British Reggae Three'. *Black Music*, September: 14–19.
May, Chris. 1977b. 'British Reggae Two'. *Black Music*, August: 18–22.
May, Chris. 1977c. 'The State of British Reggae: One – in the Belly of the Monster'. *Black Music*, July: 14–18.
McCormack, Derek P. 2013. *Refrains for Moving Bodies: Experience and Experiment in Affective Spaces*. Durham: Duke University Press.
McKittrick, Katherine and Alexander G. Weheliye. 2017. '808s and Heartbreak'. Available online: https://trueleappress.com/2017/10/12/808s-heartbreak/ (accessed 21 March 2019).
McKittrick, Katherine, ed. 2015. *Sylvia Wynter: On Being Human as Praxis*. Durham: Duke University Press.
McKittrick, Katherine. 2006. *Demonic Grounds: Black Women and Cartographies of Struggle*. Minneapolis: University of Minnesota.
McLuhan, Marshall. 1967. *The Gutenberg Galaxy*. London: Routledge & Kegan Paul.
McMillan, Michael. 2005. 'The West Indian Front Room: Memories and Impressions of Black British Homes'. Available online: https://www.youtube.com/watch?v=kRydM-pOGPg (accessed 17 July 2019).
Melville, Caspar. 2020. *It's a London Thing: How Rare Groove, Acid House and Jungle Remapped the City*. Manchester: Manchester University Press.
Mizen, Phil. 2004. *The Changing State of Youth*. Houndmills, Basingstoke: Palgrave Macmillan.
Modern Times. 1996. 'Modern Times LTJ Bukem Documentary'. Available online: https://www.youtube.com/watch?v=KrWvB6aiwFg (accessed 23 April 2018).
Moten, Fred. 2003. *In the Break: The Aesthetics of the Black Radical Tradition*. Minneapolis: University of Minnesota Press.
Nandy, Ashis. 1983. *The Intimate Enemy: Loss and Recovery of Self Under Colonialism*. Delhi; Oxford: Oxford University Press.
Natal, Bruno. 2008. *Dub Echoes*. Soul Jazz Records.
NFTR. 2015. 'Ghetts – History, P Money Clash, Kano, Movement and More'. Available online: https://www.youtube.com/watch?v=Q9sjnFvhC74 (accessed 20 January 2020).
Nicolov, Alice. 2017. 'The History of UK Pirate Radio – and Why it's Still Here'. Available online: https://www.dazeddigital.com/music/article/34394/1/pirate-radio-history-and-future (accessed 20 January 2020).

NME. 1981a. 'Directory of Sounds'. Available online: http://uncarved.org/dub/splash/directory.html (accessed 14 September 2017).

NME. 1981b. 'Interview – Sir Coxsone Outernational'. Available online: http://www.uncarved.org/dub/splash/coxsone.html (accessed 14 September 2017).

NME. 1981c. 'Shaka: Spiritual Dub Warrior'. Available online: http://uncarved.org/dub/splash/shaka.html (accessed 14 September 2017).

No Lay. 2008. 'Unorthodox Daughter'. YouTube. Available online: http://www.youtube.com/watch?v=C3Pp3V0N0hk (accessed 15 August 2012).

No Lay. 2011. 'Exclusive Interview with @HOOD2HOODtv'. Available online: https://www.youtube.com/watch?v=bGfaqGr4_VI (accessed 2 January 2020).

Paintin, William. 2016. 'Grime, Hip Hop and Technology in Urban Music Scenes'. Independent Study Project for BA Music, Goldsmiths College, London.

Palmer, Lisa. 2011. '"LADIES, A YOUR TIME NOW!" Erotic Politics, Lovers' Rock and Resistance in Britain'. *African and Black Diaspora: An International Journal*, 4, no. 2: 177–92.

Partridge, Christopher. 2010. *Dub in Babylon*. London: Equinox.

Pekar, Harvey. 2010. 'Sophisticated Basses: The Pioneering Players of Duke Ellington's Golden Years'. Available online: http://www.bassplayer.com/artists/1171/sophisticated-basses-the-pioneering-players-of-duke-ellingtons-golden-years/26190 (accessed 5 October 2017).

Pilkington, Lucy. 1994. 'All Black: Jungle Fever'. Available online: https://www.youtube.com/watch?v=cK0y9U578yk (accessed 24 April 2018).

Plummer, John. 1972. 'Handsworth and the Villa Cross Affair'. *Race Today*, 4, no. 7: 219.

Prager, Karen Jean. 1995. *The Psychology of Intimacy*. New York: Guilford Press.

Pryce, Ken. 1986. *Endless Pressure: A Study of West Indian Life-styles in Bristol*. 2nd edn. Bristol: Bristol Classical.

Putnam, Robert D. 2000. *Bowling Alone: The Collapse and Revival of American Community*. London: Simon & Schuster.

Race Today Collective. 1977. 'The Policing of Carnival'. *Race Today: The Voice of the Black Community in Britain*, 9, no. 6: 127–31.

Rashid, Naaz. 2016. *Veiled Threats: Producing the Muslim Woman in Public Policy Discourses*. Bristol: Policy Press.

Remarc and Funky Flirt. 1994. Weekend Rush.

Reynolds, Simon. 2008. *Energy Flash: A Journey Through Rave Music and Dance Culture*. New edn. London: Picador.

Riley, Mykaell. 2014. 'Bass Culture: An Alternative Soundtrack to Britishness'. In Jon Stratton and Nabeel Zuberi (eds), *Black Popular Music in Britain since 1945*, pp. 101–14. Farnham: Ashgate.

Risky Roadz. 2017. 'The Best of Risky Roadz'. Available online: https://www.youtube.com/watch?v=qw0HXLbsb74 (accessed 20 January 2020).

Roll Deep. 'One On One With Roll Deep'. Available online: https://www.youtube.com/watch?v=41cDsY88GRs (accessed 20 January 2020).

Rooper, Alison. 1996. 'Radio Renegades: UK London Pirate Radio Documentary'. Available online: https://www.youtube.com/watch?v=Hg3iZHZD7bQ (accessed 24 April 2018).

Rosso, Franco. 1981. *Babylon*. Icon Film.

Rubinstein, Daniel and Katrina Sluis. 2013. 'The Digital Image in Photographic Culture: Algorithmic Photography and the Crisis of Representation'. In Martin Lister (ed.), *The Photographic Image in Digital Culture*, pp. 22–40. 2nd edn. Abingdon: Routledge.

Ruff Sqwad. 2003. *Tings in Boots*. London: Ruff Sqwad Recordings.

Rush, Ed et al. 1993. Don FM 105.7FM.

Salewicz, Chris. 2012. 'Duke Vin: "Soundman" Who Brought Sound Systems to Britain'. Available online: https://www.independent.co.uk/news/obituaries/duke-vin-soundman-who-brought-sound-systems-to-britain-8336228.html (accessed 4 July 2019).

Sartre, Jean-Paul. 1969. *Being and Nothingness: An Essay on Phenomenological Ontology*. London: Methuen.

SBTV and All 4. 2017. 'Pirate Mentality: How Pirates Made Grime'. Available online: https://www.youtube.com/watch?v=4lSFWTx3L6Y (accessed 23 April 2018).

SBTV. 2009. 'Fem Fel | Introduces F64 to SBTV – F64 [S1.EP1]: SBTV'. Available online: https://www.youtube.com/watch?v=P22TKdMATz4 (accessed 20 January 2020).

Sedgwick, Eve Kosofsky and Adam Frank. 1995. 'Shame in the Cybernetic Fold: Reading Silvan Tomkins'. In Eve Kosofsky Sedgwick and Adam Frank (eds), *Shame and its Sisters: A Silvan Tomkins Reader*, pp. 1–28. Durham, NC; London: Duke University Press.

Seely, Rachel. 1994. 'All Junglists – a London Somet'ing Dis'. Available online: https://www.youtube.com/watch?v=9UjeXt_V3hA (accessed 5 January 2020).

Sennett, Richard. 2009. *The Craftsman*. London: Penguin.

Sexton, Richard E. and Virginia Staud Sexton. 1982. 'Intimacy: A Historical Perspective'. In Martin Fisher and George Stricker (eds.), *Intimacy*, pp. 1–20. New York: Plenum.

Seymour, Richard. 2019. *The Twittering Machine*. London: Indigo Press.

Sharma, Sanjay, et al. 1996. *Dis-orienting Rhythms: The Politics of the New Asian Dance Music*. London: Zed Books.

Shaw, Stephanie J. 2013. *W.E.B. Du Bois and the Souls of Black Folk*. Chapel Hill: University of North Carolina Press.

shouldermove. 2010. '35 Classic Jungle Drum & Bass Breaks – A to Z'. Available online: https://www.youtube.com/watch?v=HICxFSrjzN0 (accessed 20 January 2020).
Shystie. 2004. *Diamond in the Dirt*. Polydor.
Shystie. 2011. 'Ima Boss Featuring DVS'. YouTube/Iamshystie. Available online: http://www.youtube.com/watch?v=JWw2aBq5jtU (accessed 21 August 2012).
Simmel, Georg and Donald N. Levine. 1971. *On Individuality and Social Forms: Selected Writings of Georg Simmel*. Chicago; London: University of Chicago Press.
Sir Coxsone Sound. 1975. *King of the Dub Rock*. Safari.
Sir Coxsone Sound. 1982. *King of the Dub Rock Part 2*. Regal Records.
Sivanandan, Ambalavaner. 1990. 'All That Melts into Air is Solid, the Hokum of New Time'. In Ambalavaner Sivanandan (ed.), *Communities of Resistance: Writings on Black Struggles for Socialism*, pp. 19–60. London: Verso.
Skepta. 2015. 'Shutdown'. Available online: https://www.youtube.com/watch?v=MQOG5BkY2Bc (accessed 20 January 2020).
Slobodian, Quinn. 2018. *Globalists: The End of Empire and the Birth of Neoliberalism*. Cambridge, MA: Harvard University Press.
Smith, Daniel W. 1997. '"A life of pure immanence": Deleuze's "critique et clinique" project'. In Gilles Deleuze (ed.), *Essays Critical and Clinical*, pp. xi–liv. Minneapolis: University of Minnesota Press.
Smith, Matthew and Frances Porter. 2010. 'Champion Sound'. Available online: https://www.youtube.com/watch?v=kRydM-pOGPg (accessed 10 January 2020).
Solomos, John. 2003. *Race and Racism in Britain*. 3rd edn. Basingstoke: Palgrave Macmillan.
Spivak, Gayatri Chakravorty. 2008. *Other Asias*. Malden, MA; Oxford: Blackwell.
Stoler, Ann Laura. 2010. *Carnal Knowledge and Imperial Power: Race and the Intimate in Colonial Rule*. New edn, with a new preface [by the author]. Berkeley: University of California Press.
Sullivan, Paul. 2014. *Tracing the Dub Diaspora*. London: Reaktion.
Taraska, Julie. 1996. 'Invisible Jukebox: Goldie'. *The Wire*, February: 40–1.
Thompson, E. P. 1976. *William Morris: Romantic to Revolutionary*. New York: Pantheon Books.
Titley, Gavan. 2019. *Racism and the Media*. London: Sage.
Trilling, Daniel. 2018. *Lights in the Distance: Exile and Refuge at the Borders of Europe*. London: Picador.
Trower, Shelley. 2012. *Senses of Vibration: A History of the Pleasure and Pain of Sound*. New York: Continuum.

Turkle, Sherry. 2011. *Alone Together: Why We Expect More from Technology and Less from Each Other*. New York: Basic Books.
Upcoming Movement. 2010. 'UpComing MoveMent – Freestyle & Speak of CD Releases [new]'. Available online: https://www.youtube.com/watch?v=3s6uVbPB-BQ (accessed 28 January 2014).
Upcoming Movement. 2012. 'Home Town Glory'. YouTube/IceFilmsUK. Available online: http://www.youtube.com/watch?v=pzZCJ9Tuonw (accessed 16 August 2012).
Valluvan, Sivamohan, et al. 2013. 'Critical Consumers Run Riot in Manchester'. Available online: http://dx.doi.org/10.1080/14797585.2012.756245 (accessed 18 February 2013).
Valluvan, Sivamohan. 2016. 'Conviviality and Multiculture: A Post-Integration Sociology of Multi-Ethnic Interaction'. *Young*, 24, no. 3: 204–21.
Valluvan, Sivamohan. 2019. *The Clamour of Nationalism: Race and Nation in Twenty-First-Century Britain*. Manchester: Manchester University Press.
van der Hoeven, Arno. 2012. 'The Popular Music Heritage of the Dutch Pirates: Illegal Radio and Cultural Identity'. *Media, Culture & Society*, 34, no. 8: 927–43.
Various artists. 2005. *Run the Road*. 679 Recordings.
Veal, Michael E. 2007. *Dub: Soundscapes and Shattered Songs in Jamaican Reggae*. Middletown, CT: Wesleyan University Press.
Vin, Duke. 2017. 'Duke Vin Interview'. Available online: https://www.youtube.com/watch?v=yAdAbSDl7-s (accessed 17 July 2019).
Wajcman, Judy. 1991. *Feminism Confronts Technology*. Cambridge: Polity.
Weheliye, Alexander. 2005. *Phonographies: Grooves in Sonic Afro-Modernity*. Durham: Duke University Press.
Weizman, Eyal. 2007. *Hollow Land: Israel's Architecture of Occupation*. London: Verso.
White, Joy. 2017. *Urban Music and Entrepreneurship*. Abingdon: Routledge.
Whitfield, Gregory Mario. 2002. 'Aba Shanti Interview'. Available online: http://www.uncarved.org/dub/aba/aba.html (accessed 17 July 2019).
Wicke, Peter. 2016. 'The Sonic: Sound Concepts of Popular Culture'. In Jens Gerrit Papenburg and Holger Schulze (eds), *Sound as Popular Culture: A Research Companion*, pp. 23–30. Cambridge, MA: MIT Press.
Wighton, David. 1998. 'Mandelson Plans a Microchip Off the Old Block'. *Financial Times*, 23 October.
Wiley. 2004. 'Wot Do U Call It'. Available online: https://www.youtube.com/watch?v=e1YKFV45M18 (accessed 20 January 2020).
Wiley. 2010 [2001–6]. *The Matrix Instrumental*. Avalanche Music.
Wiley. 2017. *Eskiboy*. London: Windmill Books.

Williams, Raymond. 1977. *Marxism and Literature*. Oxford: Oxford University Press.
Wilner, Waren. 1982. 'Philosophical Approaches to Interpersonal Intimacy'. In Martin Fisher and George Stricker (eds), *Intimacy*, pp. 21–38. New York: Plenum.
Wolfson, Sam. 2013. 'Giggs: Prison, Police Harassment, Cancelled Tours – When Will It Stop'. Available online: https://www.theguardian.com/music/2013/oct/05/giggs-when-will-it-stop (accessed 20 January 2020).
Wolton, Alexis. 2010. 'Tortugan Tower Blocks? Pirate Signals from the Margins'. Available online: https://datacide-magazine.com/tortugan-tower-blocks-pirate-signals-from-the-margins/ (accessed 28 November 2017).
Wynter, Sylvia and Katherine McKittrick. 2015. 'Unparalleled Catastrophe for our Species? Or, to Give Humanness a Different Future: Conversations'. In Katherine McKittrick (ed.), *Sylvia Wynter: On Being Human as Praxis*, pp. 9–89. Durham: Duke University Press.
Yousef, Nancy. 2013. *Romantic Intimacy*. Stanford, CA: Stanford University Press.
Zeldin, Theodore. 1994. *An Intimate History of Humanity*. London: Sinclair-Stevenson.
Zhou, Renjie, et al. 2010. 'The Impact of YouTube Recommendation System on Video Views'. IMC '10. Melbourne, Australia.
Zuberi, Nabeel. 2014. '"New throat fe chat": The Voices and Media of MC Culture'. In Jon Stratton and Nabeel Zuberi (eds), *Black Popular Music in Britain since 1945*, pp. 185–201. Farnham: Ashgate.

INDEX

affectivity 16–18, 100–1
algorithms 88, 101
alternative
　culture 2, 5, 53, 102–3, 106, 111–19
　economy 92, 102–3 (*see also* autonomous economy)
anti-colonial
　struggle 43
　texts 108
anti-racist
　activism 56
　debate 110
　politics 112
　scholarship 109–11
apprenticeship 41, 72
authenticity 8, 70, 78, 81, 114
authoritarianism 106
autonomous economy 53, 57, 79, 91, 102–3 (*see also* alternative, economy)

Babylon 37
Barbican 89
bass 38, 41, 48, 74, 97–8, 114
Benjamin, Walter 7
Berlant, Lauren 12
Betts, Leah 63
black
　activist 36 n.11
　British 5, 28, 37
　Caribbean 29 n.3
　cultural politics 23, 43, 46, 56

cultural production 9, 43
　family 37–8
　music 8, 29, 58, 71, 79
　people 8, 33, 34
　radical aesthetics 18, 43
　US culture 4, 9
　vernacular 55, 57
　youth culture 64
black Atlantic 4, 8, 60 n.4, 94, 112 (*see also* black diasporic)
black diasporic 2, 8, 68, 112 (*see also* black Atlantic)
　experience 44
　history 53
　sound culture 1–3, 5–6, 8–9, 18, 19 n.12, 20, 22, 39, 44, 53, 57, 78, 81–2, 85, 90, 92, 97–9, 102 n.19, 103
Blaupunkt Blue Spot (*see* Blue Spot)
blues dance 46–7
Blue Spot 37
bourgeois 5, 7, 10–12, 78, 86, 107–11, 115
Bowie, David 34 n.8
Broadcasting Act (1990) 63
Bulger, Jamie 63
buzz 73 n.16, 73–5, 77, 118 (*see also* hype)

Canary Wharf 86
capitalism 5, 8, 32, 41, 53, 72, 85–6, 100, 107–12

digital 116 (*see also* prosumption)
neoliberal 108 (*see also* neoliberal)
care 117
car stereo 60
cell phones (*see* mobile phones)
Channel U 83 (*see also* Platt, Darren)
Clapton Eric 34 n.8
class 107–12
colonial 11, 63, 109
   actions 34
   governance 11 n.6
   intimacy 11
   Victoriana 38
   violence 12
colonialism 43, 107, 110 (*see also* Empire)
community 37, 64–5
Conflict video 91
Conservative 24, 33, 33 n.7, 62, 85 (*see also* Tory)
consumer culture 68, 71–2, 77, 79, 88, 90–1, 105, 116
conviviality 70, 77, 104, 115, 117 (*see also* multiculture)
Count Spinner 41
craft 39–42, 116
Criminal Justice and Public Order Act (1994) 63
crisis of capitalism 32

Debord, Guy 6
deejay 31
DeNora, Tia 16
Department of Trade and Industry 64
depth 14, 16, 24, 50–1, 76–7, 104
deterritorialization 87–9
diaspora 45–6, 113
DIY 38–9, 71–3, 89–92, 97
dominant culture 5, 111, 113

Douglass, Frederick 19 n.12
Dread, Mikey 28
dub 50–2, 115
Du Bois, W. E. B. 4
DVD 82–3, 91

education 62
Ellison, Ralph 9–10, 20–2
Empire 11, 100, 107
Eshun, Kodwo 20
Eurocentric 3, 8, 10–12, 21, 53

femininity 11–12
file sharing 83
finitude 88, 117
FM signal 77
frequency 114–15
front room 38

gender 16, 20, 101
gift 90, 117
Giggs 81
Gilbert, Jeremy 17
Gilroy, Paul 8
grime 92–6, 118
   aggression and agonism 95–6
   coldness 94–5
   criminalization 84 n.4
   dirt, soil and ground 93–4
GRM Daily 85
Grossberg, Lawrence 16–17

hype 56, 73–6, 118
hyperlocality 87–9

immediacy 76–7, 115
impermanence 76–7, 115
Internet 83–4
intimacy 3, 10–16

Jah Shaka 49–51
Jameson, Fredric 8
Johnson, Linton Kwesi 4

jungle 58–9, 63
  etymology 68
jungle pirate radio 58–9, 99
  criminalization 55, 63–4, 79
  massive 61, 65–7, 117
  studios 59
  transmission 59

Keefe, Rooney 82–3 (*see also* Risky Roadz)
kinship (*see* community)
knowledge 14–15, 43–4, 72 (*see also* wisdom)

late modernity (*see* modernity)
Lawrence, Stephen 62
Lefebrve, Henri 7
left
  critique 107–12
  scholarship 107
Levinas, Emmanuel 14–16
Lewis, Gail 13–14
Link Up TV 85
lo-fi 97
logos 6–7
love 105–6
lyricism 99, 114

Major, John 63
Mangrove Restaurant 36
Marcel, Gabriel 14–15
masculinity 21–2, 29 n.1, 68, 79, 87, 89, 114
MC culture 99, 114
McKittrick, Katherine 113, 113 n.4
middle-class 33, 37, 111 (*see also* bourgeois)
Millennium Dome 86
mobile phones 61, 61 n.6, 84–5, 89, 97
  Sony Ericsson W810 84, 89

mobility 56, 57 n. 2, 87, 98–9, 104, 114
modernity 2–9, 18, 21, 57, 79, 108, 111–13, 118
Morris, William 41–2
Moten, Fred 19 n.12
MP3 98
MTV 91
multiculture 45, 69–71, 103
multi-ethnic (*see also* conviviality)
  city 45, 69, 78, 86, 103, 118
  culture 69, 104
  genre 102
  relation 53, 56, 77, 115
  society 2, 37, 70, 102 n.19, 109
mutuality 72, 106, 116–17

nationalism 106–8, 110
neighborhood 87
neo-liberal
  capitalism 24, 33 (*see also* capitalism)
  city 86, 114
  -ism 61–4, 85–6
New Cross fire 35, 35 n.9
New Labour 85–6
NME 27, 30
No Lay 92–3, 96

patriarchy 7, 11–12, 19 n.12, 21, 79
Pay as You Go Cartel 84
phonic 18
pirate radio 57, 87
  intimacy 67
  Jungle (*see* jungle pirate radio)
  sea pirates 57–8
  terrestrial pirates 5
Platt, Darren 83
pleasure 41, 72, 74–5, 78
plenitude 88, 117
policing 35

postcode areas 87
postcolonial analysis 7–8, 14, 25, 107, 113
post-soul music 20
Powell, Enoch 34 n.7
pre-modern 8–9, 9 n.4, 112
presence 14, 42–3, 65–7, 87–9
primordial (see pre-modern)
private sphere 11–14, 19 nn.11–12, 67, 67 n.11
privatisation 42, 117
property 10–11, 14 n.8, 15–16, 72, 79, 103, 107, 110–13, 115–17
prosumption 90, 103, 117
proximity (see presence)
Pryce, Ken 46–7
psychology 13
public
   space 87
   sphere 12–13, 19 n.11, 67, 74, 98

race 1, 8, 11, 21–2, 70–1, 79, 101–2, 104 (see also racism)
racial (see racism)
racial capitalism 6, 19 n.12, 23, 25, 53, 112–13
racism 8, 11, 33–4, 62, 100, 102, 106–12, 114–15
radio 57–61 (see also pirate radio)
   intimacy debate 68
raves 61, 78
reggae sound system 5–6 (see also sound system)
relation 6, 13, 106, 112, 118
   grime 87, 90, 92
   jungle pirate radio 59–61
   reggae sound system 29
rhythmanalyst 7
Risky Roadz 82–3 (see also Keefe, Rooney)
Rock Against Racism 34 n.8
Rose, Nadia 1–2, 105–6

SBTV 85
screen 99–102
Sidewinder 82
signification 17 (see also textual culture)
Sir Coxsone Outernational Sound System 30, 39
Skepta 89
slavery 8
social media 105–6, 110
sonic 6, 8, 79, 92, 94
   screen 99–102, 114
sonic intimacy 16, 113
   of grime YouTube music videos 102–4
   of jungle pirate radio 77–9
   of sound systems 52–3
Sony Ericsson W810 84 (see also mobile phones)
soul 19
   music 18–19
sound 6, 48, 70–1
   lo-fi 97
sound system
   audience 32
   crew 30
   criminalisation 35–6
   dancehall 32
   deejay 30
   leisure 35
   relation 30–1
strategic essentialism 109–10

Tebbit, Norman 62, 62 n.8
technology 8–10, 39–42, 71–3, 89–92
   Japanese consumer technology 59–60, 60 n.4
   sound system 30
   telecommunications 61
temporality (see time)
textual culture 5, 7, 8, 16–18, 100, 113 (see also logos)

Thatcher, Margaret 33, 34 n.6
time 51–2, 76–7, 115
Tory 2, 56, 63, 68, 81, 96 (*see also* Conservative)
treble 98, 114

UK Garage 93

VHS 82
vibe 46–50, 118
vibration 49
visual culture 5, 6, 8, 18, 47–8, 69–71, 99–102, 113–15
voice
   human 99, 101, 114–15

Weheliye 8–9
welfare 61, 65, 117
whiteness 6, 95, 99–100, 103, 109
white noise 74

white working-class 108–9
wholeness 14–15, 46, 104, 118
Wiley 84, 94–5
Wireless Telegraphy Act (2006) 83
wisdom 14–15, 41, 44–5, 67–9, 89, 98, 103–4, 115–17 (*see also* knowledge)
working-class 2, 5, 28, 33–5, 52, 62, 85, 94, 107–12
   city 78, 94
   marginalization 53
   neighbourhoods 87
   performance 102
   politics 45
   sound culture 102 n.19
   struggle 43

YouTube 85, 87–92, 97, 106, 114

www.ingramcontent.com/pod-product-compliance
Lightning Source LLC
Chambersburg PA
CBHW070737230426
43669CB00014B/2488